CAMBRIDGE LIBRARY COLLECTION

Books of enduring scholarly value

Earth Sciences

In the nineteenth century, geology emerged as a distinct academic discipline. It pointed the way towards the theory of evolution, as scientists including Gideon Mantell, Adam Sedgwick, Charles Lyell and Roderick Murchison began to use the evidence of minerals, rock formations and fossils to demonstrate that the earth was older by millions of years than the conventional, Bible-based wisdom had supposed. They argued convincingly that the climate, flora and fauna of the distant past could be deduced from geological evidence. Volcanic activity, the formation of mountains, and the action of glaciers and rivers, tides and ocean currents also became better understood. This series includes landmark publications by pioneers of the modern earth sciences, who advanced the scientific understanding of our planet and the processes by which it is constantly re-shaped.

Considerations on Volcanos

George Julius Poulett Scrope (1797–1876) published *Considerations on Volcanos* in 1825. The work contains the results of his observations of volcanos in the volcanic regions of central France, Italy and Germany. It includes scientific descriptions of all volcanos in these areas, with each categorised according to its level of activity, main characteristics and geological history. Scope's work was one of the first attempts at a comprehensive theory of volcanic action and an understanding of the significance of volcanos as evidence for the earth's history. Scrope argued that volcanos should be studied in terms of known geological processes, and that 'non-catastrophic' causes should be considered to explain their formation. He argued that a gradual cooling of the earth was key to the formation of volcanos. This is a major work of nineteenth-century geology that sets out many of the principles still followed in vulcanology.

Considerations on Volcanos

Leading to the Establishment of a New Theory of the Earth

George Poulett Scrope

CAMBRIDGE UNIVERSITY PRESS

Cambridge, New York, Melbourne, Madrid, Cape Town, Singapore,
São Paolo, Delhi, Dubai, Tokyo, Mexico City

Published in the United States of America by Cambridge University Press, New York

www.cambridge.org
Information on this title: www.cambridge.org/9781108072304

This edition first published 1825
This digitally printed version 2011

ISBN 978-1-108-07230-4 Paperback

The original edition of this book contains a number of colour plates, which cannot be printed cost-effectively in the current state of technology. The colour scans will, however, be incorporated in the online version of this reissue, and in printed copies when this becomes feasible while maintaining affordable prices.

Additional resources for this publication at www.cambridge.org/9781108072304

CONSIDERATIONS

ON

VOLCANOS,

THE PROBABLE CAUSES OF THEIR PHENOMENA,
THE LAWS WHICH DETERMINE THEIR MARCH,
THE DISPOSITION OF THEIR PRODUCTS,

AND THEIR

CONNEXION WITH THE PRESENT STATE AND PAST
HISTORY OF THE GLOBE;

LEADING TO THE ESTABLISHMENT OF

A NEW THEORY OF THE EARTH.

By G. POULETT SCROPE, Esq.
SEC. GEOL. SOC.

London:
PRINTED AND PUBLISHED BY W. PHILLIPS, GEORGE YARD, LOMBARD STREET;
SOLD BY W. & C. TAIT, EDINBURGH;
AND HODGES & M'ARTHUR, DUBLIN.

1825.

PREFACE.

THE object of the following Essay is to throw some light on those phenomena which consist in the development of subterranean activity in the form of Volcanos and Earthquakes, the investigation of which appears to me of primary importance to the progress of Geological science.

The science of the History of the Globe has, I conceive, for its object—the examination of the nature of the inorganic substances observable within the limits of our planet and its atmosphere, with a view to discover its origin and the changes it has undergone.

It is thus naturally divided into two branches, viz.

I. The examination of these substances and their mutual relations.

II. The establishing conclusions from the results of this examination, as to their origin, and the changes they have undergone.

The first division is properly called *Geognosy*. The term *Geology* should of right be restricted to the second.

Geognosy embraces the subordinate sciences of Mineralogy, Chemistry, and the study of the relative positions of Rocks (*Geognosie positive* of Humboldt.)

Geology has for its business a knowledge of the processes which are in continual or occasional operation within the limits of our planet, and the application of these laws to explain the appearances discovered by our Geognostical researches, so as from these materials to deduce conclusions as to the past history of the globe.

The surface of the globe exposes to the eye of the Geognost abundant evidence of a variety of changes which appear to have succeeded one another during an incalculable lapse of time.

These changes are chiefly,

I. Variations of level between different constituent parts of the solid surface of the globe.

II. The destruction of former rocks, and their reproduction under another form.

III. The production of rocks *de novo* upon the earth's surface.

Geologists have usually had recourse for the explana- of these changes to the supposition of sundry violent and extraordinary catastrophes, cataclysms, or general revolutions having occurred in the physical state of the earth's surface.

As the idea imparted by the term Cataclysm, Catastrophe, or Revolution, is extremely vague, and may comprehend any thing you choose to imagine, it answers for the time very well as an explanation; that is, it stops further inquiry. But it has also the disadvantage of effectually stopping the advance of the science, by involving it in obscurity and confusion.

If, however, in lieu of forming guesses as to what may have been the possible causes and nature of these changes, we pursue that which I conceive the only legitimate path of geological inquiry, and begin by examining the laws of nature which are actually in force, we cannot but perceive that numerous physical phenomena are going on at this moment on the surface of the globe, by which various changes are produced in its constitution and external characters; changes extremely analogous to those of earlier date, whose nature is the main object of geological inquiry.

These processes are principally,

I. The Atmospheric phenomena.

II. The laws of the circulation and residence of Water on the exterior of the globe.

III. The action of Volcanos and Earthquakes.

The changes effected before our eyes, by the operation of these causes, in the constitution of the crust of the earth are chiefly—

I. The Destruction of Rocks.

II. The Reproduction of others.

III. Changes of Level.

IV. The Production of New Rocks from the interior of the globe upon its surface.

Changes, which in their general characters bear so strong an analogy to those which are suspected to have occurred in the earlier ages of the world's history, that, until the processes which give rise to them have been maturely studied under every shape, and then applied with strict impartiality to explain the appearances in question; and until, after a long investigation, and with the most liberal allowances for all possible variations, and an unlimited series of ages, they have been found wholly inadequate to the purpose, it would be the height of absurdity to have recourse to any gratuitous and unexampled hypothesis for the solution of these analogous facts.

The study of the processes by which these effects are at present produced on the surface of the globe forms therefore a most important, but unfortunately most neglected, branch of Geology.

It comprehends the following divisions, viz.

I. Meteorology, or the knowledge of the atmospheric phenomena, including the decomposition of rocks by the air and light, Electricity, Magnetism, &c.

II. Hydrology, or the study of the laws that regulate the conduct of water on the surface of the globe, and its mechanical effects on the solid parts of that surface.

III. That of the phenomena produced on the exterior of the globe by the development of its internal and subterranean activity.

Each of these is equally necessary for a due appreciation of the effects which such agents are capable of producing, during an unlimited series of ages, on the constitution of the globe; and at the same time offers in itself a subject of extreme interest.

The following work is an attempt to pursue the last branch of study; and, however imperfect, I am not without hope that it may be found useful in clearing up much of the obscurity and doubt which at present exists as to the nature and mode of action of the volcanic forces.

The opportunities which I have enjoyed for the prosecution of this study were sufficiently favourable to claim for me some confidence in the general facts I advance on my own authority.

My attention was first directed to the phenomena of Volcanos in the winter of 1818, which I passed within sight of Vesuvius, then in permanent eruption. In the course of the next year I visited Ætna and the Lipari Isles, particularly Stromboli, where I spent some days. After having examined the phenomena of these active volcanos, I turned my attention to the extinct vents of Italy and France; passed six months in Auvergne, the Velay, and Vivarais, contiuually occupied in geological researches, and afterwards revisited Vesuvius (which I reached just in time to witness the stupendous eruption of October 1822), examined the Ponza Isles, and all the different volcanic districts of Italy from thence to the Euganean Hills, returning to England by those of the Rhine, the North of Germany and the Eiffel.

In addition to these original sources of information, I have availed myself of many of the facts that are to be met with in the works of Humboldt, Von Buch, Beudant, Brieslak, Brocchi, Monticelli, De la Torre, Mackenzie, Bory de St. Vincent, and other writers who have published descriptions of phenomena or countries of a volcanic nature.

The short sketch of a History of the Earth with which I have been tempted to terminate this essay, may probably be considered as premature; but the conclusions to which I arrived in the body of the work, on the nature of the volcanic phenomena, unavoidably led to their generalization in this form; and since the theory is confessedly imperfect, requiring much further development, room is left for the corrections and improvements that may be hereafter suggested by other researches, hydrological and meteorological, which should perhaps have been made to precede even the rough draft of such a history.

The Appendix will be found to contain a list of all the volcanos known to be at present in habitual activity, taken in great part from that published at Paris in the Annuaire du Bureau des Longitudes, for the year 1824; to which Messrs. Humboldt and De Buch contributed.

To this list has been added brief accounts of some of their most remarkable eruptions and other contemporaneous phenomena, from the descriptions of eye-witnesses.

An article is also appended to the work, on the remarkable volcanic eruption of Xorullo in 1756, in which I have taken the liberty to dispute the inferences drawn by M. de Humboldt from the traditionary account of the phenomena, and the appearances at present visible on the spot.

I trust the theory I have ventured to put forth will be considered impartially and without prejudice; and, if so, I think it can hardly fail to satisfy a candid and enlightened reader.

For myself, when the idea was first urged upon me by the same chain of inductive evidence which is given in the work, I was incredulous; but the more I investigated its results, the more closely did they appear to correspond with, and explain the problematical appearances of the earth's crust, which have so long presented a stumbling-block to Geology.

Just as these pages are going to press, I have met with the work of Mr. Knight on a new theory of the earth, which I had not previously known. I observe that some of his ideas as to the mechanical deposition of gneiss, mica-slate, &c., and the crystallization of rocks by the admixture of an aqueous vehicle in very small proportions, correspond with mine. We differ however *toto cœlo* on most points. He supposes mountains to have been original protuberances of the granitic surface of the globe, and does not admit of the subsequent elevation of their strata; and notwithstanding this, he conceives the strata hardened by heat. He imagines that the admission of water to the metallic nucleus of the globe produces volcanos, &c.; ideas which are completely at variance with the opinions as to the nature and modes of operation of the subterranean forces, developed in the following Essay.

TABLE OF CONTENTS.

CHAP. I.—Page 1.

Descriptive Account of the Volcanic Phenomena.

CHAP. II.—Page 17.

Fxplanation of the Volcanic Phenomena.

CHAP. III.—Page 67.

Disposition of Volcanic Products on the Surface of the Globe.

CHAP. IV.—Page 85.

Laws of the Protrusion and Disposition of Lava, en masse, on the Surface of the Earth.

CHAP. V.—Page 110.

Consolidation of Lavas.

CHAP. VI.—Page 134.

Divisionary Structure assumed by Lavas on their Consolidation.

CHAP. VII.—Page 150.

Volcanic Mountains.

CHAP. VIII.—Page 171.

Subaqueous Volcanos.

CHAP. IX.—Page 187.

Systems of Volcanos.

CHAP. X.—Page 200.

Development of Subterranean Expansion in the Elevation of Strata, and Production of Continents above the Level of the Ocean.

CHAP. XI.—Page 220.

Origin of the Strata composing the Crust of the Globe, involving a New Theory of the Earth.

APPENDIX.—Page 245.

APPENDIX.—Page 245.

ON VOLCANOS, &c.

CHAP. I.

Descriptive account of the Volcanic Phenomena.

§. 1. THE action of a Volcano, in its simplest and most general form, may be described as, the rise of earthy substances in a liquefied state and at a high temperature, (Lava,*) from beneath the outer crust of the earth; accompanied by prodigious volumes of elastic fluids, which, ap-

* *Lava*, is properly any rock in a state of natural liquefaction by heat; and this term does not imply its existence in the form of a current (couleé), or any other mode of disposition. Scoria have been in the state of lava, as well as the more compact blocks. The oldest trachytes were produced from the earth as lavas. It is idle to attempt to confine the term to those rocks which exist as currents, and deny it to those which have the form of dykes, masses, &c.; since these different dispositions are often assumed by separate portions of the same continuous rock. Thus, when a current of lava has flowed over a surface in which any fissures had been previously formed, by earthquakes, or other causes, the liquid matter has occupied these cavities, producing veins, or dykes, which take off, like roots, from the overlying bed.

Those geologists who insist on confining the term *lava* to a rock in the form of a current, must, in these and similar instances, assert the vertical dykes to be a different rock from the flat bed of which they form the prolongation, and with which they are identical in mineralogical character; which would manifestly be absurd. This untenable distinction is in reality dictated by a remnant of the spirit of the Wernerian scepticism on volcanic subjects: those who have long combatted the Plutonian opinions naturally retain an extremely sensitive antipathy to the terms *lava* and *volcano*, and are anxious to confine their meaning within the narrowest possible limits.

When beaten in an argument as to facts, the necessity of a retractation, and the mortification of the *peccavi* may be often evaded by a skilful stickling about words.

Upon their consolidation, lavas become rocks of trachyte, basalt, &c. according to their mineral composition.

pearing to be evolved from the interior of the mass, burst upwards with violent successive detonations, scattering into the air, to a considerable height, numerous fragments, still in a liquid state, of the lava, through which they tear their way, together with shattered blocks of the solid pre-existing rocks, which obstructed their expansion.

§. 2. These phenomena, which are currently designated as *Volcanic Eruptions*, are not confined to a few particular localities, but take place, either continually, or at intervals of greater or less duration, from numerous points, in all quarters of the globe.

The known active volcanos, or *habitual vents of volcanic energy*, have been reckoned to amount to nearly 200 in number.

This calculation must be considered as greatly below the amount of those that really exist on the globe.

I. Because a large proportion of the earth's surface remains unknown to us, and it is probable that volcanos exist in that part as well as in the rest.

II. Because these phenomena are most frequent and energetic exactly in those quarters of the globe which have been least explored; viz. in the interior of N. and S. America, and amongst the innumerable archipelagos of the great Pacific Ocean. There is therefore great reason to suppose, that, when these districts are better known, we shall find them to contain more volcanos than those of which we have as yet received intelligence.

III. Because the intervals of rest, which occur between the eruptions of a volcano, are sometimes of such long duration, that all accounts of their former occurrence are forgotten, and the volcanic character of the mountain remains unknown, until a fresh eruption proclaims the continued activity of its subterranean focus.

IV. Because there is good reason to believe that very many volcanic vents exist at the bottom of the sea; few of which can be brought under our observation before the orifices of eruption have been raised above the level of its surface.

Taking all these circumstances into consideration, it is surely allowable to suppose that we are not acquainted with more than one half of the occasionally active volcanos existing on the surface of the solid crust of the globe.

In running over the catalogues* of active volcanos which have been framed from the accounts of such phenomena reported by eye-witnesses on different parts of the globe, we find them irregularly scattered over the whole surface of our planet, occurring indifferently in all latitudes, and under every meridian; sometimes detached singly, and at a considerable distance from each other; but generally either concentrated into close groups, or forming a connected linear chain or series; in some very rare instances seated in the interior of a continent: usually, however, rising as insular mountains from the depths of the ocean, or at very little distance from its borders, upon a maritime coast.

§. 3. Amongst the circumstances which must most materially characterise any individual development of the volcanic forces, are those which relate to the situation of the point on which the eruption breaks through the solid crust of the earth: and most particularly whether this point is placed above or below the surface of any of the great bodies of water, which cover so large a portion of our planet.

In the first of these cases, the eruption takes place in open air; in the second, in water; and the different density of these media must considerably modify the nature of the phenomena, and the conduct of the substances, whether gaseous or solid, produced from the interior of the earth. We shall therefore consider separately these different kinds of eruptions, of which the first may be called *sub-aerial*, the second *sub-aqueous*.

The latter class of eruptions, though we have reason to believe them by no means of rare occurrence, can at least be but very rarely observed; and their phenomena, even when the attention of those who chance to be passing near the spot, is called to them, can be only partially and imperfectly perceivable. The water may be observed to be more or less agitated, heated, and discoloured, and to be traversed by the rising column of gaseous fluids emitted below, and even by jets of the fragmentary matters they carry up with them; but it is only when the accumulation of the solid substances pro-

---

* Vide Histoire des Volcans de M. Ordinaire—Brieslak, Institutions Geologiques, tom. iii. Annuaire du Bureau des Longitudes, An. 1824. A catalogue compiled from these and other sources is given in the Appendix, No. 1.

truded from below, have at length elevated the vent above the water level, and the phenomena consequently take place in open air, that they become more immediately liable to direct observation, and they then enter into the first class of eruptions—*sub-aerial.*

It is therefore to this class principally that we must direct our attention, in studying the laws of volcanic action.

Phenomena of Sub-aerial Eruptions.

§. 4. The character of a volcanic eruption taking place in open air will vary considerably, according as it proceeds from a *new* or an *habitual* vent. Since, in the first case, it must force a passage through rocks previously uninfluenced by catastrophes of this nature, and in which an immense resistance is, in all probability, opposed to the rise of the lava and aeriform fluids, by the cohesive force, as well as the weight, of solid and undisturbed strata of various kinds.

In the second case, on the contrary, the eruption has been preceded by a long series of similar explosions from the same point, an issue has been previously broken through the continuous strata of other formations, a road opened for the occasional discharge of superabundant activity, and the only resistance to be overcome will consist in the greater or less obstruction of the vent thus established, by the accumulation within and above it, of the solid substances produced by prior eruptions.

§. 5. It is obvious that an eruption occuring from a vent newly opened on the surface of a continent, must be modified, both in its circumstances and results, by the nature and disposition of the solid rocks it traverses. Examples of such phenomena are of very rare occurrence; indeed few or none have been recorded that may be strictly reckoned such.

The eruption of the Monte Nuovo, in the bay of Baiæ, can scarcely be allowed to bear this character, since it occurred in the midst of the volcanic soil of the Phlegrean fields. The same must be said of the eruptions on the Mexican plateau which produced Yorullo and five other volcanic cones (as described by Humboldt), where the country around exhibited previous traces of volcanic action. These examples, therefore, belong rather to the class of subsidiary apertures to habitual, or previously active volcanos.

But in those districts where the volcanic activity appears to have been but recently extinguished, if it is in reality yet quenched for ever, we have many remarkable instances of the occurrence of single eruptions on fresh and isolated points.

The high plateau of granite, for example, which rises to the west of the Limagne, in the French province of Auvergne, and separates the waters of the Allier and Sioule, presents a chain of above seventy distinct cones, produced in this manner. And upon the same linear band, prolonged through the provinces of the Velais and Vivarais, are to be seen a string of similar cones, amounting in number to more than 200. Each of these is visibly the produce, usually of one, at the utmost, of three or four eruptions; and, throughout the whole line, the ejections of each separate eruption may be distinguished, and their number counted, with more or less accuracy; so that this country presents a field for the study of the different modifications of volcanic eruptions, infinitely superior to that which can be offered within any limited period, by any one active volcano existing.

The district of the Eiffel, and that lying on the left bank of the Rhine, above Andernach, afford similar observations; and though on a less extensive scale, and generally of a less recent date, are remarkably instructive, from the diversity of the rocks they have produced, from the varied nature of the strata broken through, and from the peculiar deposits of an igneo-aqueous character* by which they are accompanied, and the origin of which their state of great preservation sufficiently discloses.

§. 6. Still, however interesting the deductions we are enabled to draw from the study of rocks produced by volcanos only recently extinct, it is from positive observations on the volcanic phenomena in full activity, that we must expect to derive the soundest knowledge of their nature and mode of operation. And for this purpose it is necessary to direct our attention more particularly to those habitual vents of volcanic enery, from which eruptions take place, either unceasingly, or at intervals of greater or less duration, and which by the accumulation of their products have universally been surrounded by a mass of considerable size, constituting a *volcanic mountain.*

§. 7. A rapid review of the information we possess on the phenomena of the different known volcanic mountains, or

---

* *Trass*

habitual sources of volcanic products on the globe, leads to the conclusion that there exists the most complete irregularity with respect to the periods and intensity of their activity.

Some volcanos are for ever in a state of incessant eruption; some, on the contrary, remain for centuries in a condition of total outward inertness, and return again to the same state of apparent extinction after a single vivid eruption of a day's duration; while others exhibit an infinite variety of phases intermediate between the extremes of vivacity and sluggishness.

But upon a closer examination, it is easy to discern that, amidst all this primâ facie irregularity, a certain degree of order and harmony exists; a general correspondence of consequences to preceding circumstances is perceivable; and we are led to suspect that the apparent anomalies are but oscillations about a fixed course of action, determined by the peculiar condition of the individual volcano, and as subject to the influence of a general law, as any other of the great recurring phenomena of nature.

§. 8. The various conditions hinted at above, as exhibited by the different active volcanos, may be distinguished into three general classes, viz.

I. In which the volcano exists incessantly in outward eruption—Phase of permanent eruption.

II. In which eruptions, rarely of any excessive violence, and which continue in a comparatively tranquil manner for a considerable time, alternate frequently with brief intervals of repose.—Phase of moderate activity.

III. In which eruptive paroxysms, generally of intense energy, alternate with lengthened periods of complete external inertness.—Phase of prolonged intermittences.

Class I.

Phase of Permanent Eruption.

From a variety of circumstances it may be suspected that this condition can be but of rare occurrence, and indeed we know of but one well authenticated instance in which it is exhibited. The volcano of Stromboli, one of the Lipari islands, is always in a state of ceaseless eruption, and there is good reason to believe this condition to have lasted without intermission during the last 2000 years at least. The aeriform fluids here continue to escape by successive explosions from the same habitual *vent*, apparently as fast as they are generated within the subjacent mass of ebullient lava, which never overflows the lip of the orifice, and is rarely, if ever, emitted otherwise than as projected scoria.*

The volcano in the lake of Nicaragua, called by our sailors, the Devil's mouth, is likewise represented as existing in the state of continual activity—incandescent scoria are constantly thrown up from it in jets: but, as at Stromboli, the lava is seldom, or never, discharged in any quantity.

Perhaps amongst the numerous volcanos of the Oriental Archipelagos, other similar examples may be hereafter discovered.

Class II.

Phase of Moderate Activity.

§.10 This condition is common to a great number of volcanos, and particularly exemplified in those which, from their proximity to the seats of civilization and science, have been as yet almost exclusively studied.

It is characterised by the alternation of minor eruptions, or periods of prolonged external activity, with brief intervals of quiet.

Such was the condition of Vesuvius, from the commencement of the present century, up to October, 1822: during

* I was assured by the inhabitants of the island, who are almost solely fishermen, and therefore have the volcano constantly under their observation from the sea—that the intensity of its eruptive violence is much greater in winter than in summer, and usually encreases in proportion to the storminess of the season, so much so that they augur fair or foul weather by the state of the volcano. If their accounts are to be believed, the abrupt face of the cone, which shelves down almost perpendicularly from the volcanic orifice to the sea, is occasionally split open in the storms of winter, and discharges a current of lava into the sea, by which the water is heated and discoloured, and fishes destroyed so as to be cast on shore ready boiled. The volcanic phenomena of this island are remarkably interesting. I shall hereafter return to the consideration of the circumstances which may be the cause of the extraordinry equilibrium that, in this rare instance, seems to be perpetually maintained between the volcanic force, and the resistance opposed to it, by the weight of the elevated lava and the obstruction of its vent. Sir W. Hamilton, in his "Campi Phlegræi," also mentions, and gives an engraving of the lateral escape of a current of lava from the side of this singular cone.

which time the volcano has remained frequently in eruption, for the space of five or six months together, discharging jets of scoria and sand, from some temporary orifice at the summit of the cone; while diminutive streams of lava, sometimes two or three in number at a time, welled out with the tranquillity of a water-spring, from openings in its side.

These periods of activity were succeeded by intervals of rest, lasting perhaps some few months, till interrupted by the recommencement of eruptions, resembling the preceding ones, but always from new orifices, broken through the summit of the cone; the prior issues being apparently choked by the consolidation of the lava they produced, and the weight of the small cones raised by the loose ejections.

In October, 1822, this state of things terminated in an eruptive paroxysm of great violence, which, in all probability (like that of 1794), will have effected a change in the condition of the volcano, and bring it into the third phase, or that of prolonged intermittences.

The same volcano appears to have existed in a smilar phase of continued activity, from 1767 to 1779, when a violent eruption, analogous to that of 1822, put a period for a time to the continuance of the phenomena, and reduced the volcano again to the third phase in which it remained till about 1803.

The actual state of Ætna offers another instance of the volcanic phasis under review.

During the last 20 years four principal eruptions have taken place, viz. in 1805, 1809, 1811—12, and 1819, each of which produced a considerable quantity of lava; but the intervals between these epochs of remarkable excitation, were not without various minor phenomena attesting the continued activity of the focus. Frequent earthquakes were felt, not only by the inhabitants of the mountain's flanks, but often through the whole island; one of these shocks (16th February, 1810) is said even to have extended its influence as far as Cyprus. Smoke was almost continually emitted by the crater; accompanied occasionally by detonations, and probably by jets of scoria, since the appearance of flames so often mentioned by Signor Gemmelaro,* was in all likelihood caused by the light of the jets which took place at the bottom of the deep crater, reflected from the impending cloud of smoke and vapour.

Many of the volcanos studding the Pacific, as well as some of that great train of active volcanic vents which stretches in

* Vide Annals of Philosophy, 1822.

a sinuous line from the N. extremity of the Peninsula of Kamskatchka through the Japan, Loo-choo, and Philippine isles, the Moluccas, and Java, into Sumatra, and the island chain of the Andamans, &c. seem, from the meagre information we possess concerning them, to exist in this state of prolonged but moderate activity; since the same volcanos, are seen to be constantly in eruption by the crews of the different vessels that navigate these Archipelagos.

This is probably the condition of the volcanos of Barren island, of Arjuna in Java,* of the small island between Timor and Ceram, of that of New Britain, seen successively by Dampier, D'Entrecasteaux, Lemaire, and Schouten, in eruption; of that of Tanna in the Archipelago of the New Hebrides, seen in activity by Cook, D'Entrecasteaux, and Forster; of the Peak of Ternate in the Moluccas; of those of Mutova and Tharma in the Kurile islands, and some others in Japan, and the Aleutian Isles.

Amongst the American volcanoes we may instance as existing in this phase that of Popocatepetl in Mexico, which has been continually active since the period of the conquest of Mexico, (see Humboldt) and that of Sangay in Quito, which has been in incessant activity since the year 1728.

The volcano of the Isle de Bourbon offers another remarkable example of the phase under consideration.—From accounts given by M. Hubert, who is described by Bory de St. Vincent, to have directed his attention to its phenomena ever since the year 1760, we know it to have existed during the last 60 years in a continual state of moderate activity, vomiting lava, at least twice in every year—eight of its currents produced in this space of time have reached the sea, and, with the others, cover a wide slope called *Le Pays brulè*, of a horrible aspect, almost entirely destitute of vegetation, uninhabited, and, from the peculiar glassy asperity of the scoriform surfaces of the currents, nearly impassable.

Class III.

Phase of Prolonged Intermittences.

§. 11. This last class of phases, viz. that of prolonged intermittences, is by far the most usual condition amongst the

---

* Sir S. Raffles's Java.

active volcanos of the globe. It has been described above as characterised by the rare occurrence of eruptive paroxysms, generally of intense energy and brief duration, separated by very long intervals of complete outward tranquillity.

The epochs of sudden and violent eruption almost peculiar to this phase, as being the most striking and terrific of all the volcanic phenomena, are those of which we have the most frequent and copious relations.

Few persons are induced to direct any particular attention to the appearances exhibited by the generality of volcanos during their periods of continual moderate activity, which, as well from their inferior grandeur, as from the constancy with which they take place, pass almost unnoticed.

But the occurrence of one of the greater paroxysms necessarily attracts the observation of the most careless spectator.

The stupendous and terrific character of these catastrophes, the rarity of their display, and the dreadful extent of injury often resulting from them to the lives and property of the inhabitants of the surrounding country, make them a subject of general remark and relation, during, and long after the period of their development.

Hence accounts of such volcanic eruptions find a place in the earliest annals of history:—play their part occasionally in the fabulous mythology of still remoter ages: and form a not unfrequent source of sublime imagery to the poets of antiquity.

§. 12. When we compare together all the accounts of such occurrences, observed in every quarter of the globe, and at distant periods, we cannot but be struck by the excessive similarity of the facts and appearances recounted; nay more, when due allowance is made for the effects of terror upon the minds of ignorant, and perhaps superstitious, observers, for the universal proneness to exaggeration of the marvellous, the want of scientific language, and the errors necessarily incidental to the relations of inexperienced persons, it is impossible not to recognise a complete uniformity, and even identity, in the main phenomena; no further discrepancies existing, than what are fairly referable to the modifications produced by local accidents, or by differences in the intensity of volcanic force developed, and in the mineral quality of the substances elevated.*

* For some remarks on the only exception which I can find to this general assertion, (that of *Jorullo*) see Appendix, No. 2.

The following is a brief sketch of the circumstances, which universally appear to characterise these great eruptions, or volcanic paroxysms.

They are usually preceded by earthquakes more or less violent, extensive, frequent, and prolonged; which are obviously caused by the efforts of the lava, swelling on all sides from the encreasing elasticity of the aeriform fluids it contains, to force a passage through the superincumbent rocks. Repeated loud subterranean detonations are heard, resembling, so as frequently to be mistaken for, the firing of heavy artillery, or the rolling of musketry; according to its intensity.

These sounds are proved, by the immense distance to which they are propagated, and with a rapidity wholly out of proportion to their loudness near the spot from which they proceed, to be conveyed not by the air alone, but chiefly by the solid strata of the earth.

Often, it is said, the state of the atmosphere assumes a peculiar character, offering an unusual closeness, stillness, and pressure.

These threatening indications of an approaching crisis are prolonged for a greater or less time, and are accompanied by the disappearance of springs, the drying up of wells, and such accidents as the cracking, splitting, and heaving, of the substructure of the mountain, must naturally occasion. The eruption begins; generally with one tremendous burst, which appears to shake the mountain from its foundations. Explosions of aeriform fluids, each producing a loud detonation, and gradually encreasing in violence, succeed one another, with great rapidity, from the orifice of eruption, which is in almost every instance the central vent, or crater, of the mountain. This vent has usually been obstructed, during a long preceding period of repose, by the ruins of its sides, brought down by the wasting influence of the weather, and the shocks of earthquakes, or by the ejections of previous minor eruptions.

The elastic fluids therefore, in their rapid escape, project vertically upwards these loose accumulated matters, and the fragments of the more solid rocks, through which they have forced a passage.

The violence and rapid repetition of these projections, to which the same fragments are exposed upon falling again towards the orifice, reduce them to such tenuity that they are carried upwards by, and remain suspended for a time in, the heated clouds of aqueous vapour which are dicharged, at the same time, in prodigious volumes, from the volcanic aperture.

The rise of these vapours, thus mingled with pulverulent

matter, produces the appearance of a high column of thick smoke, based on the edges of the crater, and appearing from a distance to consist of a mass of innumerable globular clouds, pressing on each other, and incessantly urged upwards by the continued explosions. At a certain height, determined of course by its relations of density with the atmosphere, this column dilates horizontally, and (unless driven in any particular direction by aerial currents) spreads on all sides, into a dark and turbid circular cloud. In very favourable atmospheric circumstances, the cloud with the supporting column has the figure of an immense umbrella, or of the Italian pine, to which Pliny the younger compared that of the eruption of Vesuvius, in A. D. 79, and which was accurately reproduced in October, 1822. Forked and branching lightnings of great beauty are continually darted from different parts of the cloud, but principally its borders. Its continual encrease soon hides the light of day from the districts situated below it, and the gradual precipitation of the sand and ashes it contains, and which fall as the velocity of its progress is diminished, contributes to envelope the atmosphere in gloom, and adds to the consternation of the inhabitants of the vicinity.

Meantime the lava boils up the chimney of the volcano. The elastic fluids, by which it is traversed, rend and carry upward portions of its surface, as they explode from it, and form a continual fiery fountain of still liquid and incandescent fragments, which from the velocity of their motion, present an appearance at a distance that has frequently been mistaken for flame. The internal column of lava continuing to rise, it finds an issue, at length, either over the lowest lip of the crater, or from some crevice forced through the side, perhaps even at the foot of the mountain, from whence it flows in torrents. By night, the running lava appears at a white heat wherever the liquid interior of the current is visible; but as upon contact with the air its surface is instantaneously congealed into a thick scoriform crust, the general tint of the outside is a glowing red, which gradually darkens as the solidified coating encreases in thickness.

During day the lava is almost concealed from view by the torrents of aqueous vapour which rise from its whole surface in immense volumes, and unite themselves to the clouds of similar nature that hang over the mountain.

In some cases, no absolute escape of lava, in streams, takes place, scoriæ alone being projected.

In all cases where lava is emitted, its protrusion marks the crisis of the eruption; which usually attains the maximum of its violence a day or two after its commencement. The stopping of the lava in the same manner indicates the termination of the crisis, but, by no means, of the eruption itself. The gaseous explosions continue, with immense and scarcely diminished energy.

At length they cease to throw up liquid or red-hot scoriæ; the fragments projected are either blocks of older rocks, or consolidated scoriæ. By degrees, these fragments, most of which fall back into the crater, become more and more comminuted by the immense trituration they sustain in the process of repeated projection and fall; till at length clouds of sand alone and ashes, reduced in the end to an extraordinary degree of fineness, are carried upwards by the eructations of the aeriform fluids.

These explosions gradually decrease in violence, appearing to be stifled by the accumulations of finely pulverised fragments, which occupy the volcanic vent and impede their expansion.

The column of ashes projected becomes gradually shorter, until at length all struggle seems to cease: no further explosions are heard; the eruption has terminated; usually however not for many days, or even weeks, after attaining its maximum of violence. Soon the crumbling in of the crater's sides choaks up still further the volcanic orifice, and conceals it from view. An interval of quiet then commences, of a protracted duration, forming the other characteristic of the phase we are considering.

These tremendous demonstrations of volcanic energy are always accompanied or followed by more or less violent meteoric phenomena; sometimes equally terrific and destructive with the former; the atmosphere appearing to share in the convulsion which agitates the earth. The summit of the volcanic mountain necessarily attracts and condenses the volumes of aqueous vapours which have risen from the volcanic orifice, and the lava emitted; and hence a fall of rain takes place in prodigious quantity on its sides and base, producing torrents, which, carrying with them the ashes, sand, scoriæ, and fragments, with which the slopes are strewed, rush, as deluges of liquid mud, towards the plains or vallies below, and cover them with vast deposits of volcanic alluvium.

As instances of Volcanic Paroxysms, amongst many on record we may point out those of

Vesuvius, in the years A. D. 79, 203, 472, 512, 685, 993, 1036, 1139, 1306, 1631, 1760, 1794, and 1822.

Ætna, in 1169, 1329, 1535, (this eruption lasted two years with terrific violence, and occurred after a quiescence of nearly a century,) 1669, 1693—4, 1780, and 1800.
Teneriffe in 1704, and 1797—8.
San Georgio, one of the Azores, in 1808.
Palma, one of the Canary isles, in 1558, 1646, and 1677.
Lancerote, belonging to the same group, in 1730.
Kattlagaia Jokul, in Iceland, A. D. 1755, which lasted a year.
Skaptar Jokul, in 1783. In fact, all the volcanos of Iceland appear to exist solely in this phase.*

§. 13. Such are the phenomena which characterise the display of the volcanic forces from an habitual vent, at the moments of paroxysm. These efforts are generally preceded, and more constantly *followed*, by long periods of complete tranquillity; the energies of the volcano seeming to be exhausted, for a time, by the violence of their development.

The duration of this quiescent interval is of very unequal continuance, extending even occasionally to centuries; and thus, it frequently happens, that the superficial scoriæ of the cone, and its internal cavity, or crater, become so far decomposed, as to afford a soil in which various vegetables find sustenance. All appearances of igneous action are effaced; forests grow up and decay; and cultivation is carried on, upon a surface, destined perhaps to be blown to atoms, and scattered to the winds, when the crisis arrives for the renewal of the volcanic phenomena. Thus during the quiescent interval, between the eruptions of 1139 and 1306, the whole surface of Vesuvius was in cultivation, and pools of water and chesnut groves occupied the sides and bottom of the crater; as is at present the case with so many of the extinct craters of Ætna, Auvergne, the Vivarais, &c.

In general, after the cessation of a paroxysm, many *fumarole*, or emanations of vapour evolve themselves from the lava currents which were then produced, as well as from the bottom of the crater. These vapours are at first almost wholly aqueous, but at a later period generally contain some mineral acids, and deposit various saline incrustations at the mouths of the fumarole. When the acidity of the vapours is in excess, and

* In the Appendix will be found some brief descriptions of individual occurrences of this character.

their production continued for a long period, they effect a very considerable degree of decomposition on the exposed parts of the rocks against which they act,* and the crater of a volcano in this condition is said to pass into the state of a Solfatara or Souffriere. Such is the actual condition of the Solfatara near Pozzuoli, the Souffrieres of St. Vincent, Guadaloupe, and St. Lucia, in the Caribbee Isles; of the great central crater of the Peak of Teneriffe; of the craters of Milo in the Archipelago; of Volcano, one of the Lipari Islands; of Crabla in Iceland, and of Tanna, one of the New Hebrides, according to Dr. Forster.

This condition is by no means a proof of the complete extinction of the volcano, as was proved by the eruptions from the Souffriere of St. Vincent in 1812; which had been completely tranquil since 1119; and by a similar renewal of activity from those of Volcano in 1786, and of Guadaloupe in 1778, 1797, and 1812.

§. 14. All the volcanos of the Atlantic, whether in Iceland, the Azores, Canaries, Cape Verd, or Caribbee Isles, appear at present to exist in this third phase of prolonged intermittences. A great proportion of those which stud the Cordilleras of the two American continents, and nearly all which occur in Sumatra, Java, the Moluccas, Japan, Kamskatka, and the numerous Archipelagos of the Pacific, belong to the same class.

Throughout these two great volcanic trains (which perhaps in reality form but one) we hear of terrific eruptions occasionally breaking out from mountains which were not previously suspected to be of volcanic nature, or in which the accounts of former catastrophes of this sort existed but as vague traditionary fables.

§. 15. It has been thought right thus to distinguish these three classes of modifications under which the phenomena of a volcanic vent habitually show themselves, in order to simplify the study of their nature and mode of operation. But it must be recollected that this distinction is purely artificial; and many volcanos will of course be found to exist in intermediate conditions, partaking of the characters of more than one class. The same volcanic vent also occasionally passes from one phase into another. Thus Vesuvius, which appears

---

* For the effect of these decomposing vapours, see a succeeding Chapter.

to have generally existed in the third phase, as far as can be deduced from the imperfect accounts which are preserved of its phenomena, and of which, of course, from the reasons mentioned above, only the most terrible and rarest paroxysms will have been noted, seems to have continued during a great part of the 17th century, in the second phase, frequent eruptions having been mentioned between the years 1660 and 1694; while again after a quiescence of ten years, it returned in 1804 into the scond phase, and was almost constantly active until 1822. Ætna in the same manner appears to have passed into the second phase towards the beginning of the 17th century; since which epoch more than 40 eruptions are recorded, with but one quiescent interval of any considerable duration, viz. from 1702 to 1755. Some of these eruptions however have been decided paroxysms, particularly those of 1669 and 1787.

These changes have probably often taken place in other instances; many volcanic mountains bearing marks of having experienced such varieties of condition. It will be seen hereafter that they are to be expected from what we know of the nature of the volcanic forces.

CHAP. II.

Theory of the Volcanic Phenomena.

§. I. THERE can be little doubt that the main agent in all these stupendous phenomena, the power that breaks through the solid strata of the earth's surface, elevates lavas to the summits of lofty mountains, and launches still higher into the air the shattered fragments of the rocks that obstructed its efforts, consists in the expansive force of elastic fluids struggling to effect their escape from the interior of a subterranean mass of *lava*, or earths in a state of liquefaction at an intense heat.

It is also scarcely to be questioned that these aeriform fluids are generated in the lava by means of its exposure to the intense heat, which produces its liquidity. In other words, that this substance exists in a state of either temporary or continual *ebullition*.

The observation made by Spallanzani, in 1788, on Stromboli, first exhibited the nature of volcanic agency in its true light. In the spring of 1819, I had an opportunity of verifying on the spot the accuracy of the circumstances related by the Italian professor.

The phenomena which he observed still take place in a precisely similar manner.

The actual aperture of this volcano, at the bottom of its semicircular crater, is completely commanded by a neighbouring point of rock, of rather perilous access, from whence the surface of a body of melted lava, at a brilliant white heat, may be seen alternately rising and falling within the chasm which forms the vent of the volcano. At its maximum of elevation one or more immense bubbles seem to form on the surface of the lava, and rapidly swelling, explode with a loud detonation. This explosion drives upwards a shower of liquid lava, that, cooling rapidly in the air, falls in the form of scoriæ. The surface of the lava is in turn depressed, and sinks about 20 feet, but is propelled again upwards, in a few moments, by the rise of fresh bubbles, or volumes of elastic fluids, which escape in a similar manner; and it is evidently this incessant evolution of aeriform substances in vast quantities, which preserves the lava invariably at so great an elevation, within the cone of Stromboli, and constitutes the permanent phenomena of its eruptions.

In this instance there evidently exists within and below the cone of Stromboli, a mass of lava, of unknown dimensions, permanently liquid, at an intense temperature, and continually traversed by successive volumes of aeriform fluids, which escape from its surface—thus presenting exactly all the characters of a *liquid in constant ebullition.*

The phenomena of other volcanos which take place only at intervals of greater or less duration do not appear to differ from those of Stromboli in the nature of their agents. Thus, during the eruption of Vesuvius in 1753, those who ventured to the summit of the cone observed jets of liquid scoriæ thrown up successively from the surface of a mass of lava, at a brilliant heat, which occupied the bottom of the crater, and conducted itself exactly in the manner of a liquid in ebullition.

Spallanzani remarked a similar appearance within the crater of Ætna in 1788. The volcano of the Isle of Bourbon presents another parallel fact. Bory de St. Vincent, who twice visited the active crater of this cone, and passed a whole night upon its borders, describes it as filled with a body of liquid lava, apparently at an intense heat, but covered by a thin and cracked pellicle or crust, except in the centre, where it was completely incandescent and continued alternately swelling upwards, and falling again, after giving vent to a jet of liquid lava; the oscillatory motion, thus communicated to the body of lava, produced a series of concentric waves that wrinkled its surface.

There exists indeed no account of the occurrence of any considerable subaerial eruption, unaccompanied by the vertical projection of fragments of liquid and incandescent lava, (scoria or pumice); which are sufficient, though no currents should be produced, to prove that the elastic fluids explode from the interior of a mass of that nature.

In the case of minor eruptions from volcanic mountains of large size having a central crater of great depth, it must of course occasionally happen, that the incandescent scoriæ thrown up from the bottom of this cavity are invisible to any spectator placed below the summit of the mountain, (and few have the courage or curiosity to seek this situation at such a moment). In these cases, however, the shock of the detonations is felt, and their report heard, by the inhabitants of the mountain's flanks; while the brilliant light of the jets of scoriæ, reflected in flashes from the clouds of aqueous vapour and ashes impending over the cone, produces that appearance which is so repeatedly described under the erroneous appellation of *flames*, in the accounts given of volcanic eruptions by inexperienced or unscientific witnesses.

§. 2. We must therefore allow the phenomena of volcanos during their periods of eruption, to prove the existence at such times, below every vent, of a subterranean mass of liquified earths (lava), of indeterminate extent, and at an intense temperature, traversed, more or less freely, by volumes of elastic fluids, which burst from its surface, and rise rapidly into the upper regions of the atmosphere. Such a mass of lava obviously conducts itself exactly in the manner of a liquid in *ebullition*.

§. 3. But the liquidity of lava, even at the moment of its protrusion from the orifice of a volcano, is extremely imperfect and of a very peculiar character. There is no reason for supposing it to be a homogeneous liquid like water, or the metals and earths in a state of complete fusion. On the contrary, every circumstance in its conduct and composition shows its analogy to those compound liquids, such as mud, paste, milk, blood, honey, &c., which consist of solid particles deriving a certain freedom of motion amongst one another from their intimate admixture, in greater or less proportion, with one or more perfect fluids, which act as their *vehicle*.

The ideas currently entertained of the liquidity and complete fusion of lava, are nothing more than vague impressions suggested by the appearances it presents of a high temperature and considerable mobility of parts. But accurate observations by no means bear out this conclusion as to its real nature.

During some of the tranquil and continuous eruptions, which have been already described as constituting one of the habitual phases of the volcanic phenomena, opportunities have frequently been afforded of approaching and examining a stream of lava at its very source in some crevice of a volcanic mountain. At the moment of emission, it appears a substance at a brilliant white heat, radiating very little caloric, compared with that which proceeds from any metal when glowing with similar intensity. Its liquidity, when greatest, does not appear to exceed that of honey; but is generally so imperfect, that considerable pressure is required to cause a pointed stick, or blunt rod, to penetrate its surface. This surface is observed to consolidate, almost instantaneously, upon exposure, at the same time emitting a considerable quanity of vapour. The superficial crust thus formed cracks and splits tumultuously in all directions, and fresh vapours escape from the crevices. Its brilliancy of glow is at the same time rapidly dimmed, and its colour passes through red to black. The crust suddenly hardened in this manner, has the same aspect and character as the scoriæ which are thrown up into the air by the gaseous explosions from the main vent of the volcano.

The very small quantity of heat radiated by the incandescent lava, is attested by the little effect it produces on a thermometer, held at the distance of a few feet from its source, and by the slowness and difficulty with which glass melts or iron is heated to redness, even by contact with it. The observation has been continually made by all who have observed the production of lava, and impressed Dolomieu in particular with the idea that its temperature is much lower than is supposed, and that it owes its liquidity to some other principle than caloric. The low radiating powers of the incandescent lava may be in part perhaps attributed to the rapid congelation of its surface; the crust thus formed being known to conduct caloric with such extreme difficulty and slowness, that a stream of lava still flowing on in the interior may be approached within the distance of a few feet, without injury to the exposed parts of the person, and is often traversed on foot.

But the *thickness* of this crust, and the distance to which the process of congelation rapidly progresses inwardly, must prevent our contenting ourselves with this explanation. We are therefore reduced to suppose with Dolomieu, that the lava is *not* in a state of complete fusion, but owes the mobility of its particles to some other circumstance than their complete combination with caloric, and can part with this quality without any intense manifestation of heat.

§. 4. This idea as to the nature of the liquidity of lava is strongly confirmed, by reference to its composition and characters on consolidation.

In fact if a portion of lava rock of any quality be reduced to *complete fusion* by artificial heat, and then exposed to contact with the air, it hardens into a glass without any traces of lithoidal or crystalline texture; whereas the same substance, when naturally propelled from the earth in a liquid state, and consolidated under precisely the same circumstances, almost invariably presents a lithoidal or crystalline structure. With the exception indeed of a few comparatively rare masses which possess a vitreous or resinous texture, and in which alone any approach to perfect fusion can be conceded, *all* the rocks produced in this manner appear to consist of an intimate aggregation of crystals, varying in size, and frequently so minute that their facets are not distinguishable with the naked eye; by which the rock acquires an aspect of perfect compactness. Even their fine grained parts are, however, found by the lens, and by the aid of mechanical analysis, to consist equally of a crystalline aggregation of the same minerals that compose the coarser parts, or that are disseminated in large and visible crystals through the apparently homogeneous base. It has been supposed that this peculiarity of texture may be owing to various circumstances of pressure or slowness of refrigeration, accompanying the consolidation of the rock in natural lavas; but so far from this being the case, it is a remarkable fact that the scoriform crust which rapidly congeals on the surface of a current by exposure to the open air, is found to possess the same crystalline texture, and to contain the same proportion of larger or minute crystals, as the innermost parts of the current produced *en masse* by the same eruption, and consolidated in the slowest and most tranquil manner. The same thing exactly is observable of the *scoriæ* which congeal even still more rapidly by sudden projection into the air from the liquid mass of lava within the vent; and even a fragment withdrawn with a hooked stick from the incandescent margin of a current; and which is consolidated instantaneously before the eyes of an observer, offers precisely the same characters of texture and composition. It is impossible to conceive that the process of crystallization can take place during this instantaneous and tumultuous refrigeration, particularly with respect to the large and regular crystals.

We are driven therefore to conclude that the crystalline texture of most lavas is not owing to any circumstance accompanying their consolidation, but that the greater number, if not all, of its component and imbedded cystals existed ready formed antecedently to its emission from the volcanic orifice; and, also, which is a necessary consequence of this observation, that this substance *is seldom in a state of igneous fusion, but owes its partial and imperfect liquidity to some other cause, by which a certain mobility is communicated to the solid crystalline particles of which it consists.*

§. 5. No species of liquidity can exist independent of caloric. But the quantity of caloric that produces the liquidity of any substance, may be either combined in a uniform manner and in equal proportions with all its molecules; which is the case when a body is in complete *fusion;* or it may be merely combined with one ingredient of the mass; or with a certain portion of the constituent particles, the remainder existing in a solid state, but acquiring a freedom of motion from being suspended in the surrounding fluid.

Thus water is an example of complete fluidity (or fusion), all its particles being combined with an equal dose of caloric. But if a quantity of clay be diluted in water, the mass will still retain a more or less imperfect liquidity, the particles of clay remaining solid, but acquiring mobility from their suspension in the fluid.

In the same manner the liquidity of lava, when it issues from the earth, is by no means any proof of its fusion, but may be caused by the suspension of its crystalline particles in any accompanying fluid, aeriform or liquid. The considerations specified above have already led to the conclusion that its liquidity is really of this latter kind. The degree of its liquidity will then depend, not upon the quantity of caloric with which it is combined, but on the quantity of fluid which it contains in proportion to the size and form of the solid particles; and its consolidation will be effected, not by parting with its caloric, but by the escape or condensation of the fluid that occasioned the mobility of these solid particles. This is in exact conformity with the conduct of lava on exposure to the air. It radiates very little caloric, but discharges a great quantity of *elastic vapour* from its surface, which is instantly hardened. This elastic vapour, then, is, in all probability, the fluid to which it owed its previous liquidity; the vehicle in which its crystalline particles were suspended, and the escape of which, from the exposed surface of the lava, allows of the aggregation of its particles, and the induration of the mass they make up. But this induration must be accompanied by a diminution in bulk, or contraction, which will split the superficial crust, and produce in it a number of fissures perpendicular to the surface. These, again, by allowing the escape of more elastic vapour from the inner parts of the lava to which they penetrate, propagate the consolidation inwardly, and give rise in turn to other fissures, by which the process is still further advanced. In this manner (in reality) the superficial coating of a mass of lava, almost immediately after it issues into open air, *is* split and cracked in all directions, and the violence with which the torrents of vapour escape from these fissures, elevates into every variety of position, the ragged and scoriform fragments of the crust; which is still further broken up by the swelling or sinking of the stream of still liquid lava, that flows on beneath it, as under an arched tunnel or gutter, in obedience to the inclination of the ground; and acquires a similar scoriform coating upon every fresh point of surface which it successively exposes.

§. 6. But of what nature is the aeriform fluid which escapes so rapidly from the exposed surfaces of lava; and the loss of which causes its sudden consolidation? There need be no doubt, at all events, that it is the same fluid which causes the expansion, intumescence, and rise of this substance through the volcanic aperture, until, in most cases, it overflows the lip of this orifice; and of which large volumes struggling rapidly upwards through the column of liquefied earths, that fills this vent, escape with violent bursts from its surface, producing jets of liquid lava and red hot scoriæ.

The results of all the experiments and observations that have been made during volcanic eruptions, lead to the conclusion that the elastic fluid which plays so important a part in the phenomena of volcanos is no other than aqueous vapour, or *steam*.

The fluids that are disengaged in the greatest abundance from the superficial crevices of a current of lava,* while still visibly incandescent in its interior, have been often made the subject of direct experiment, and they appear to be *purely aqueous vapour;* or if any other principle is present, it is generally in too minute a proportion to be discovered by our means of analysis. As the lava cools further, the admixture of other elements with the steam is perceptible, though in very small proportion. These are principally the sulphuric or muriatic acids, or naphtha; and at the same time sulphur, muriate of soda, muriate of ammonia, and other salts, are deposited, as *sublimations*, round the mouth of the fissure.† It

---

* See Brieslak, Daubuisson, Humboldt, Menard de la Groye, Monticelli, &c.

† May not these acids, &c. be generated by the slow filtering of steam through narrow crevices of the lava at an intense temperature, just as steam

is possible that some permanent gasses are occasionally discharged in company with the steam, but such have been rarely, if ever, detected. The emanations of carbonic acid gas and sulphuretted hydrogen, frequent in volcanic districts, have rarely been observed to proceed from the heated lava.

This fact, that the great mass of elastic fluid discharged from a volcanic vent in eruption, consists of aqueous vapour, is with equal evidence attested by its uniformly collecting into thick and heavy clouds above the mountain; which, unless dispersed by violent winds, soon fall by further condensation in violent and destructive torrents of rain, upon the sides of the volcanic hill, and the surrounding country.

The vapours continually evolved by Stromboli, form a permanent cloud, of a white or greyish colour; which, when the air is calm, remains stationary above the peak, or encircles it with a wreath of mist, or falls in light showers, according to the density and temperature of the atmosphere: when the air is in motion it streams down the direction of the wind, like an endless ribbon, or, more exactly like the track of mingled smoke and vapour left by a steam-vessel moving against the wind. This cloud does not consist of ashes, or finely comminuted sand (for the fragmentary ejection of this volcano are heavy and coarse, and fall immediately on their projection), but, as far as can be perceived, purely of *aqueous vapour*, presenting the same conduct and character as those produced by evaporation from the surrounding sea.

The existence of water in lava may seem, at first, a strange and almost incredible notion; but the late experiments of the Honourable Mr. Knox,† as well as those performed by Spallanzani, have proved that *all the rocks* of volcanic origin contain a certain proportion of water in intimate, though apparently mechanical, union with their constituent minerals; and the water, which so frequently occurs in the cellular cavities of these rocks, might have prepared us to expect the fact. Mr. Knox, it appears, found likewise more or less of naphtha or bituminous matter, in almost all the rocks which he analyzed. This volatile ingredient escaped Spallanzani, who, however, detected muriatic acid in some of the lavas, obsi-

---

passed through a heated gun-barrel is decomposed? The oxygen may unite with the sulphur of the lava to form sulphuric acid; the hydrogen with the chlorine, or carbon, to form muriatic acid, and naphtha. These will be volatilized as fast as formed, and will rise with the remaining steam, and be condensed at the mouth of the fissure.

† Transactions of Royal Society, 1824.

dians, and pumice of Lipari. These results accord entirely with numerous observations made during the production of lava from a volcano, when the smell both of bitumen and muriatic acid is often strongly sensible.*

It is sufficient for our present purpose to have fully recognised the existence of water in lavas; nor is it by any means necessary to account for its occurrence there. Indeed this stage of our inquiries is too early for the examination of this question, which must be deferred to a later occasion.

§. 7. It appears then, from what has been stated above,

1. That the intumescence and rise of lava, within the vent of a volcano, is owing to the expansive force of an elastic fluid generated in its interior.
2. That the violent rise and escape of volumes of this fluid through the columns of lava elevated within the vent, produce the jets of lava drops and scoriæ, which constitute the principal fragmentary ejections of a volcano.
3. That the further rapid escape of this same fluid from the surface of lava, when exposed to the open air, effects its instant consolidation.
4. From these and other considerations, it appears highly probable, that it is the intimate combination of this fluid with the solid crystalline particles of the lava, which occasions their mobility, and the consequent liquidity of the mass.
5. That there is every reason to believe this fluid to be no other than the vapour of water intimately combined with the mineral constituents of the lava, and volatilized by the intense temperature to which it is exposed when circumstances occur which permit its expansion.

When therefore I speak of the ebullition of lava, I must be understood to mean the vaporization of the water, or part of the water, contained in close union with its solid crystalline particles. In the same manner as wet sand, paste, milk, soup, &c. may be made to boil, though it is only the water they contain which is really vaporized.

---

* Menard de la Groye, Humboldt, Monticelli, &c.

§. 8. Let us examine the effect of any variations in the circumstances of temperature and pressure on a substance of this nature, viz. a solid crystalline rock containing a certain proportion of water intimately combined between the crystalline molecules, and already possessing a high temperature, more than equal to the vaporization of the water under the pressure of the atmosphere alone. It is obvious that either the augmentation of temperature, the pressure remaining fixed, or the diminution of the pressure, the temperature being constant, must alike have the effect, sooner or later, of producing the commencement of ebullition in the confined water. But the vaporization of the smallest part of this fluid has the immediate effect of proportionately diminishing the temperature of the surrounding mass, by the quantity of caloric absorbed by it, (become latent,) while at the same time the immense expansive force of the vapour thus generated, re-acting from the surfaces which confine the mass of lava, must produce a great augmentation of the pressure it sustains. Hence, unless a still further reduction of pressure, or increase of temperature, takes place, a very small quantity of the water can be vaporized; and, on the whole, the conversion of water into steam in any part of the lava, will be proportionate to the reduction of the pressure sustained by this part, below the degree which prevents ebullition at the existing temperature; or, to the increase of the temperature above what is the boiling point at the existing pressure.

But since the water is merely mechanically combined with the constituents of the lava; that is, interposed between the laminæ of its component crystals, the vaporization of any part of this water must affect a proportionate separation of the crystalline laminæ between which the moisture was lodged.

The effect of this separation is, 1. The more or less complete *disaggregation* of the component crystals. 2. The *intumescence* of the lava, by the bulk of the vapour generated; and its acquisition of an expansive force equal to the sum of the elasticities of all the portions of vapours generated; by which it presses against the surfaces that resist its dilatation; and, 3. The communication of a more or less imperfect *liquidity* to the mass, resulting from the mobility and elasticity of the vapour generated between the neighbouring crystalline particles. The degree of liquidity thus communicated to the lava, must be determined by the average size and form of the component crystals or solid particles, and the quantity or volume of the vapour which acts as their vehicle.

But the volume of the vapour generated on any point will, we have seen above, depend on the ratio of the temperature and pressure. Therefore, under the same circumstances of pressure and temperature, the liquidity of the lava will vary directly with the degree of comminution of the component crystalline particles, (their form being supposed the same, or nearly so).

It is obvious that, at a certain ratio between the volume of vapour generated between the interstices of the solid particles, and their size, the mobility they acquire will be sufficient to allow of the escape of the surplus of vapour from these interstices, and its agglomeration into distinct parcels by the force of its mutual affinity. The proportion of vapour thus *liberated*, or set free, to that which remains *fixed*, on the point where it was generated between the neighbouring crystals, must vary directly with the *comminution of the solid particles*, or *fineness of grain* in the lava. But here another force acts upon these parcels of liberated vapour ; viz. the difference between their *specific gravity* and that of the liquid by which they are enveloped: by this difference they are urged to rise to the upper part of the liquid mass.

This force must in every situation vary in intensity exactly with the specific gravity of the solid particles of the lava alone, since the density of any parcel of *free* elastic vapour must be equal to that of the *fixed* vapour which occasions the mobility of the surrounding liquid.

§. 9. From these considerations, therefore, we derive the following conclusions :

1. The volume of vapour generated by any change in the circumstances of a lava, depends on the ratio of the temperature to the force of compression.
2. The intumescence or dilatation of the lava, varying with the volume of vapour produced, must follow the same law.
3. The proportion of the vapour generated, which is *liberated* and unites into parcels, to that which remains fixed, will vary with the comminution of grain.
4. The force with which these parcels, or *bubbles*, tend to traverse the liquid lava and rise to its surface, varies directly with the specific gravity of the solid lava rock, and the degree to which they are enabled to obey this impulse must depend on the liquidity of the mass

immediately above them. But the liquidity and specific gravity of a fluid determines its degree of *fluidity* (i. e. *the facility with which it yields to the impulse of gravity)*, and hence the force with which the bubbles of vapour rise and escape from a lava, varies directly with its *fluidity*.

It is evident that the crystals of which any solid mass of lava is composed, are more or less comminuted, or reduced in size, by the process of ebullition. In the first place, the generation of an elastic fluid between their plates or particles, *disaggregates* them more or less. When this *disaggregation* has proceeded to a certain extent, that is, when the volume of vapour generated has reached a certain relation to the size of the crystal, it escapes from the interstices of the contiguous crystalline particles where it was generated, proportionally disturbing their arrangement, and the relative position of their poles, and consequently *disintegrating* the crystal. Secondly, this *disintegration* is carried still further by the friction of the crystals against each other, produced by the intumescence of the lava, and still more so by the mechanical disturbance and trituration, as it were, caused by the rapid and violent rise of the parcels of elastic fluid through the upper mass of lava, when the circumstances of liquidity and specific gravity permit their escape upwards. But the volume of vapour, and the consequent intumescence of the lava, depends on the ratio of the pressure and temperature; consequently, under similar circumstances of temperature and pressure, the disintegration of the crystals of lava will vary directly with its specific gravity, by which the latter triturating force is determined.

§. 10. Having thus far considered the effect of an increase of temperature, or a diminution of pressure, on a mass of lava under such circumstances, let us examine what will follow from the reverse; namely, an increase of pressure, or a diminution of temperature. Upon the solid lava, it is clear, no corresponding change will be produced; but every diminution of temperature, or increase of pressure, on a mass, or a part of the mass, liquified in the manner stated above, must occasion the condensation of a part of the vapour which produces its liquidity, and so far tend to effect its *reconsolidation*.

The vapour that remains *fixed*, is, by either of these circumstances, when in sufficient force, recondensed into water, and the enclosing crystal *reaggregated*, resuming its former integrity and solidity. The vapour that has become *free*, and occupies separate parcels disseminated through the mass, is also proportionately condensed, and the disintegrated crystals are aggregated together in a confused and irregular manner (the relative position of their poles having been disturbed) into a *solid rock*, which will be more or less porous, in proportion to the space occupied by the parcels of liberated fluid. The absolute escape of the vapour, by the exposure of any part of the lava to a rarer medium, will produce exactly the same effect in the reaggregation of the crystals, and the consequent consolidation of the rock, by removing all impediment to the action of the attractive force of the solid particles.

§. 11. It has been seen, that the vaporization of the water contained in a lava, which constitutes that process of ebullition, to which are attributed the phenomena of volcanic eruptions, may be equally produced either, first, by the reduction of the pressure upon it without any accession of caloric, or, secondly, by the accession of caloric, and consequent increase of its temperature, the pressure remaining fixed.

§. 12. Now let us imagine the existence on any point near the surface of the globe, of a solid body of lava, of indefinite horizontal and vertical dimensions, at an elevated temperature, and cut off from all communication with the external atmosphere, or ocean, by overlying strata of solid rock.

The pressure sustained by this mass is either wholly null, or consists of the reaction of the expansive force of the elastic vapours it contains, against the surfaces by which it is enclosed.

It is obvious that no change can take place in this pressure so long as the temperature remains fixed, and the confining surfaces, equally stable; nor can we conceive how any variation can be produced in the stability of those boundaries, while the temperature, and consequently the expansive force of the confined mass, remains fixed. To effect any change in this state of things, we must therefore have recourse to the supposition of an accession of caloric to the lava. This supposition, so far from being gratuitous or unwarranted, is supported and confirmed with the strongest evidence by other considerations. Thus we know that a great quantity of caloric is continually passing off from every active volcanic vent in combination with the immense volumes of heated vapour and incandescent lavas, emitted from it. Unless the loss of caloric thus sustained, were made up by the accession of an equal, or nearly equal, quantity from below, it would be impossible for the mass of lava beneath a volcanic vent to continue in eruption, as it is frequently known to do, for centuries; or to renew its eruptions, when they had ceased. The phenomena of thermal springs suggests the same remark.

The observations lately made as to the temperature of mines, which *encreases with their depth,* lead to the conclusion that the interior of the globe, at no great vertical distance, is at an intense temperature. This internal accumulation of caloric must be continually endeavouring to put itself in equilibrio, by passing from the centre towards the circumference, wherever the conducting powers of the substances enveloping the globe, permit the transmission most readily into external space. But the conducting powers of the solid rocks which compose the outer crust of the earth, are necessarily imperfect in the highest degree, from their want of density and frequently porous structure.

What transmission of caloric can take place through the thick formations of secondary limestones, clays, shales, and sandstones?

The crystalline, and particularly the compact granitoidal rocks, seem, however, far better adapted for this purpose; and caloric would no doubt be propagated through them more readily than the former class, or through the schists, and laminar rocks.

§. 13. From these considerations it is rendered probable that a continual supply of caloric passes off from the interior of the globe towards its circumference, wherever its transmission is facilitated or permitted by the conducting powers of the intervening substances, or by the temporary opening of vents for its more free escape.

If, as I think it will appear, the phenomena of volcanos, under all their various phases of action and quiescence, together with their accompaniments of earthquakes, &c. &c. and perhaps many of the more ambiguous and obscure indications of congenerous causes visible in the constitution of the globe's surface, can be accounted for in the simplest and most satisfactory manner, according to well-known principles of physics, by this single assumption of the exposure of subterranean masses of crystalline rocks, which we know to exist, to a continual accession of caloric from below, which we have the strongest reasons for presuming a priori—in this event we shall be bound by common sense and the simplest rules of induction to accept this hypothesis with the utmost confidence, and it would be the height of irrationality and scepticism to refuse our acquiescence in it.

§. 14. Allowing therefore this inferior source of caloric, and supposing it to be transmitted to a subterranean mass of lava, more rapidly than it can pass off to the outside of the globe, through the solid crust of overlying rocks, in consequence of their inferior density and conducting powers, it is obvious that the caloric will be concentrated in the lava, and continually augment its temperature, particularly that of the lower strata which are the nearest to the source of caloric. This increase of temperature generates an increase in the expansive force of the contained vapour, the reaction of which from the confining surfaces, so long as they are absolutely resistant, occasions a corresponding increase of pressure, and this equal augmentation of the opposite forces of expansion and repression will proceed even in a more rapid ratio than the increase of temperature; since it appears from all the experiments hitherto made on substances of every kind, both solid and liquid, that the dilatation positively produced in them, by equal increments of temperature, or sensible heat, progressively increases as the temperature itself rises. It is clear that under these circumstances, the general expansive force of the confined mass of lava increases in an accelerated ratio, with no other conceivable limit than the yielding of the surfaces which impede its dilatation. But since the powers of resistance of the solid overlying rocks, occupying the surface of the earth, are necessarily limited, the moment will arrive when they are forced to give way more or less, and the first effect of the sudden expansion thus afforded to the highly compressed elastic fluids contained in the lava, is the immediate diminution of the pressure which consisted in the reaction of their elastic force. The consequence must be the generation of fresh vapour by a rapid, violent, and transient ebullition,* which ceases entirely as soon as the absorption of caloric by the vapour newly generated, and by the expansion of the preexisting vapour, has lowered the temperature of the liquid to the boiling point corresponding to the remaining pressure.

Since the accession of caloric takes place by our assumption from *below*, the temperature of the mass will be unequal

---

* Exactly similar to what takes place in the Papin's digester; or in the apparatus for heating water red-hot, when the valve or stop-cock is opened.

throughout, diminishing more or less gradually from below upwards. But the temperature and consequently the expansive force being greatest in the lowest strata into which we may suppose the body of lava divided, the superior strata will, as long as they retain any degree of liquidity or mobility of parts, be subjected to a pressure out of proportion to their expansive force, from their position between the inferior strata, in which this force is in excess, and the overlying rocks; consequently the vapours contained in these upper strata must be gradually condensed, and their liquidity at length entirely destroyed, the process of complete consolidation progressing from *above downwards*, as the excess of expansion in the lower strata is augmented.

§. 15. Whenever the overlying rocks yield in any degree to the general expansive force of the mass, the consequent ebullition will take place *first*, and *with the greatest violence* where the expansive force is highest, that is, in the *lower* strata. The sudden dilatation of these inferior strata must forcibly elevate the upper *solidified* parts, as it were, *en masse*; the pressure they sustain between the expansion of the lower part, and the weight and cohesion of the solid rocks above them, suffering them to preserve the water they enclose unvaporized, notwithstanding their intense heat.

This forcible elevation of solid rocks by a violent expansion, taking effect at a considerable depth, cannot occur without considerable rupture and dislocation.

Every such solution of continuity, suddenly produced in the rocks composing the earth's crust, must create a jarring shock, or momentary vibration, in them, which will be propagated along the prolongation of each rocky bed, with more or less violence in proportion to its solidity. The strata which are merely in contact with those that are thus affected, will share in the vibration, only in an inferior degree.

These shocks, felt with more or less intensity on the surface of the globe, according to the above conditions, and accompanied generally by a sensation of heaving, or elevation of the ground under our feet, are the phenomena to which we give the name of earthquakes; none of which take place without being accompanied by a certain, though often unappreciable elevation of the surface of the globe.

§. 16. Fissures must be created by these disruptions, both in the superficial rocks of low temperature which overlie the mass of lava, and in the upper parts of this crystalline mass itself, solidified, as we have seen above, by the superior expansive force of the lower parts. Of these crevices some must be supposed to open *downwards*, towards the confined mass of lava, and others *outwardly*; the accompanying figure will best illustrate this position.

Fig. 1.

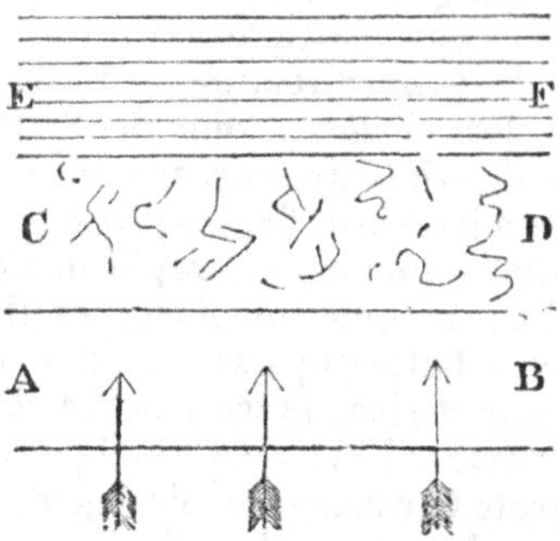

If A B C D represent the subterranean mass of lava, confined by the overlying strata E F, the accession of caloric to the lower part of the mass, viz. A B, so far increases its expansive force as to consolidate the upper mass C D (by condensing its enclosed vapour;) and by the continuation of this process, the general expansive force, acting from below upon the overlying rocks E F, at last becomes superior to their powers of resistance, and they yield more or less to the dilatation of the lower strata of lava *a b* (as in fig. 2); the fissures

Fig. 2.

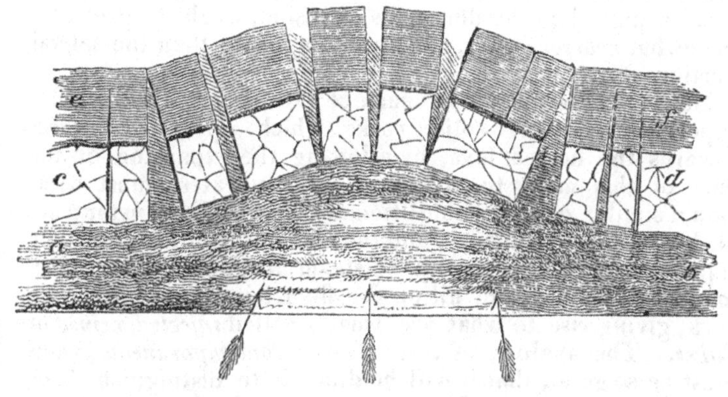

broken towards the centre or convex part of the space elevated will open outwardly, those towards the limits of this space will open downwards.

Such in fact is the natural and constant effect of any forcible elevation of a solid crust by an impulse from below.

The distinction between these two species of fissures is

important, and will be found perhaps to lead to interesting results. The lava of the upper strata *c d*, though solid, is at an intense temperature; therefore the production of any fissure in it must permit and produce the instantaneous expansion of the lava forming its sides, by the vaporization of its contained water. A rapid ebullition and intumescence, occurring in the substance of these sides, must immediately fill the fissure with liquid lava.

But the direction of these fissures will be in general more or less *vertical*; coincident with that of the expansive force by which they were occasioned; and, since the temperature of the lava increases from above downwards, the intumescence will be most violent and rapid in the lower part of the fissure, and thus a rapid movement of the liquefied lava will take place from below upwards. The friction which accompanies this violent rise, as well as the disaggregative force of the steam generated between the laminæ of the crystals, must disintegrate to a considerable degree the crystals of the ebullient lava; and hence we should expect that these *veins*,* if no further change takes place in their nature, and they become displayed by subsequent circumstances to view, will appear of a *finer grain* than the rock which they traverse. A less strongly marked difference is to be expected also in the texture of the vein itself, of which the central part, having suffered less of friction than the sides during its intrusion, ought to present a somewhat *coarser grain*, and *larger crystals*, than the lateral portions.

But in the same manner each of the fissures in the cooler superficial crust of solid rocks which opens *downwards* towards the heated lava, by allowing the expansion of the part of this mass to which it communicates, must occasion a similar effect. Ebullition will take place instantaneously with equal violence, and the intumescent lava suddenly liquefied by the generation of steam throughout its tissue, must rapidly enter and fill up the fissure in the neighbouring rock, giving rise to what are usually called *injected veins*, or *dykes*. The analogy of these to the *contemporaneous* veins must be so great, that it will be difficult to distinguish them, except in the case where the rock which they traverse is wholly of a *different mineral composition* from the veins, as when trap dykes traverse limestone, &c. Where the difference consists merely in the size of the grain, no positive

* "Contemporaneous" veins?

conclusion can be drawn as to the vein being of the first or the second class.

But the lava which fills these fissures of either kind, will speedily be reconsolidated; the injected veins with the greatest rapidity, from their contact with the cooler sides of the fissure they fill; and *both* by their exposure at a diminished temperature to the expansive force of the lava with which they communicate, and which immediately puts itself in equilibrio with the force of pressure on every point.

In this manner the overlying crust of rocks, which was considerably weakened, perhaps by the rupture and displacement of its parts, is again repaired; the fractures being cemented by the solidification of the lava injected into them.

§. 17. The expansive force of the heated lava having thus partially satisfied itself, an interval of tranquillity succeeds; until the continued accession of caloric from below again raises its temperature, which had fallen in consequence of its ebullition; and the same process recommences, the expansive force and the pressure again increasing together, until the cohesion and weight of the overlying rocks are once more overcome, and another successful effort takes place, accompanied by the circumstances, of earthquakes, and the formation of veins, already described.

It is thus that the overlying rocks which form the superficial crust of the globe must be progressively elevated more and more, by the successive dilatations of the inferior lava, wherever a subterranean body of this substance exists.

The extent of surface thus affected will depend upon the dimensions of the confined mass of lava, upon the vertical depth at which the expansion takes place, and upon the nature and disposition of the rocks composing the surface.

§. 18. At every crisis of this character, the crust is more or less fractured, and partly repaired by the injection of lava into those cracks *which open downwards.* But the fissures broken through towards the centre of the space thus elevated, open *outwardly*, towards the atmosphere; and diminish in width towards their lower extremity—these therefore will remain more or less open, and effectually weaken that part of the crust of rocks across which they are broken. The subsequent expansive efforts will therefore take effect most powerfully upon these points, widening and extending these fissures, some of which will probably at length be prolonged across the whole external crust, into the intensely heated crystalline rocks (or lava) below; by which the pressure upon it at the lower extremity of the fissure is suddenly reduced to the weight of the atmosphere alone, (supposing as we do for the present that the fissure thus formed opens above the level of the sea, or of any stationary mass of water.)

In this, as in the former case, ebullition must instantly commence. The lava at the lower extremity of the fissure is rapidly liquefied by the vaporization of its contained water, and this liquefaction is speedily propagated, both downwards and laterally, within the mass of heated rock; the intumescent lava, as it liquefies, rising rapidly up the fissure, and part of the elastic fluids generated within it escaping with still greater rapidity, and driving upwards the fragments torn from its sides. The violence and duration of the ebullition, it is obvious, must wholly depend upon the ratio in which the process of dilatation and liquefaction is propagated inwardly, through the lava, from the point at which it commences at the lower extremity of the fissure; and this ratio must be proportioned to that in which the intumescent matter is enable to rise up the fissure.

The circumstances which determine the rate of this rise are, 1. The tension, or expansive force of the confined vapours, which is the same as their temperature. 2. The width of the fissure, or *aperture* of *escape*. 3. The superficial or *external* pressure, whether atmospheric alone, or other. The rapidity of the progress of dilatation towards the interior of the heated mass of lava, and consequently the violence of the process of expansion, evidently varies directly with the two first of these terms and inversely with the last. If any two of these terms be supposed fixed, the energy of the expansion will vary with the third. Therefore, under equal circumstances of temperature and external pressure, the violence of the ebullition varies directly with the size of the fissure.

In the generality of instances it must happen that the fissure, thus broken through the superincumbent rocks, is extremely narrow, irregular, and intricate; and the distance great from the external surface of these rocks to the focus of ebullition; this process will therefore be proportionably slow and powerless, so that the refrigeration of the rising lava by contact with the sides of the fissure, together with the weight of the column of lava elevated, and the obstruction created by the fall of fragments broken off by the aeriform fluids that precede the lava in its rise, must speedily stifle the ebullition, by restoring the preponderance to the forces of repression.

The lava, so consolidated, will seal up the lower part of the fissure; and the only distinction between this case and that of the injected veins mentioned above, (§. 16.) is, that here a certain proportion of the vapour first generated will probably succeed in effecting its absolute escape into the outer atmosphere, before the fissure is finally closed. It is highly probable that the vapours, clouds of smoke, and momentary projections of fragments, that are occasionally observed to be discharged during violent earthquakes from some of the crevices which they occasion, owe their origin to a subterranean action of this nature, which may be considered as an abortive eruption, no more prolonged or more violent manifestations of subteranean activity being visible.

§. 19. But whenever the fissure broken through the overlying crust of rocks by any effort of the force of subterranean expansion, is of considerable width, and consequently produces an ebullition of such energy that the intumescence of the lava overcomes the obstacles to its rise, and is enabled to reach the mouth or upper extremity of the fissure upon any one point, *then* all the characteristic phenomena of *volcanic eruption* are manifestated on the external surface of the globe, in the more or less violent and prolonged discharge of elastic vapours and ejected fragments from an orifice in its crust.

§. 20. It may be as well to investigate more in detail the laws by which the violence and duration of the eruption must be determined.

The expansive force of a molecule of the contained water at any point to which the process of vaporization reaches, varies of course with its temperature.

The degree to which this force is developed will vary inversely with the impediments that repress its full expansion.

These are,

I. The *external* pressure; atmospheric, oceanic, or other.

II. The weight of the column of liquefied lava it supports.

III. The reaction of the vapour already generated from the confining surfaces. If these are absolutely resistant on every point except at the fissure, this reaction will vary inversely with the width of the vent.

The sum of these forces constitutes the general force of repression.

The first element of this force is wholly dependent on external circumstances.

The second varies with the specific gravity of the lava, and acts upon any point in direct proportion to its vertical distance from the highest surface of the liquefied lava. It is therefore augmented on the lava in *the focus, or seat of expansion*, in proportion as the lava rises up the fissure.

The third element of the repressive force varies, as has been said, inversely with the width of the vent, and acts upon any point in proportion to its distance from the exposed surface of the lava, measured along the line of pressure, whatever its direction.

It is seen that the two latter forces of repression encrease rapidly towards the interior of the lava, in all directions from its exposed surface, where they are both null. Consequently, if the temperature be supposed uniform throughout, the propagation of the expansion inwardly, and therefore the violence of its outward development, must proceed in a retarded ratio, decreasing more or less rapidly from its commencement: the expansive force being the same at all the points which the process successively reaches, while the repressive force is continually increasing with the distance from the exposed surface of the lava. The points at which all expansion is ultimately subdued by the predominance of the latter force, would be the limits of the space through which the dilatation is propagated.

In this manner, in reality, the progress of the dilatation in a *lateral* direction must take place in a retarded ratio, and be completely checked at a certain horizontal distance from the vent, since each horizontal stratum of lava is supposed uniform in temperature.

But we have seen that the temperature of this confined mass of heated crystalline rock increases more or less from above downwards; consequently the expansive force increases in the same direction; and if we could suppose it possible that this increase could exceed or keep pace with the rapid increase of pressure in that direction, occasioned by the weight of the supported column of liquid, no limit would exist to the progress of the expansion downwards, in a uniform, or even an accelerated ratio. But the extremely rapid augmentation of temperature, in a vertical direction, required for this, is wholly improbable; and we must conclude that, in every case, a limit exists to the inward propagation of expansion in a vertical, as well as a lateral, direction; a limit at which the repressive force, compounded of the weight of the supported column of liquid—the reaction of the vapours generated—and the external pressure, completely prevents any vaporization from taking place. The process of dilatation having attained these boundaries, the limits of the dilated mass would thenceforward remain fixed, unless the equilibrium is destroyed by some further change taking place in the circumstances that influence the energy of the opposite forces.

It is obvious that the more rapid the increase of temperature downwards, the longer will be the vertical dimensions of the dilated mass, compared to its horizontal extent; and the more will the violence of the eruption be prolonged; and, on the other hand, the more uniform the temperature of the successive strata, into which we may imagine the crystalline mass divided, the shorter will be the duration of the eruption.

Again, the greater the specific gravity of the lava, the less will be the vertical depth to which the dilatation is propagated, and the less rapid its development.

§. 21. These considerations lead to the establishment of the following general propositions; viz.

1. The energy of the eruption, or absolute quantum of outward expansion which takes place in given times, will vary *directly* with the temperature, or tension, of the lava to which the fissure communicates, and the width of this aperture; and *inversely* with the thickness of solid and refrigerated rocks across which the vent is broken, the specific gravity of the lava, and the external pressure.
2. The duration of the eruption will vary *directly* with the rate at which the temperature increases downwards, and *inversely* with the specific gravity of the lava, and width of the vent.
3. The sum of expansion produced during any finite eruption, as well as the volume of the dilated mass, will be directly proportioned to the energy and duration of the eruption.

§. 22. If we suppose the dimensions of the fissure of escape, or vent, and the external pressure to remain constant, it is obvious that the development of the expansive force will be most violent at its commencement, and decrease in energy by degrees; dying away entirely when the powers of repression and expansion had put themselves completely in equilibrio throughout every point of the expanded space. But if the external circumstances by which the force of repression is modified, are subject to change during the development of the expansive force, the ratio of that process must vary likewise, and inversely to the force of repression. Now, in all cases of volcanic eruption, it necessarily happens that the force of repression is rendered, in this manner, more or less variable by contemporaneous alterations, either

1. In the dimensions of the vent; or
2. In the external pressure or resistance to its expansion, acting upon the surface of the lava within the vent.

I. The intense energy of the dilatation that suddenly takes place at the extremity of any newly-formed fissure of escape, and the abrasive force of the dilated matters (particularly of the elastic fluids as they struggle violently upwards through this aperture) must more or less fracture and break up its sides, and thereby enlarge its dimensions.

The extent of this enlargement will depend,

1. On the violence with which the expansive force is at first developed, and therefore *(ceteris paribus)* on the temperature or tension of the confined lava.
2. On the degree of solidity of the rocks composing the sides of the fissure, which may yield more or less readily to the explosive violence of the escaping matters.
3. On the absolute depth from the surface of the earth, of the focus of expansion to which the fissure extends.

In proportion as the aperture is thus enlarged, the sum of the repressive forces is diminished, and the ratio of dilatation consequently accelerated. The energy of the expansive action—the violence of the intumescence of the lava, and the quantity protruded up the vent in given times,—are thus progressively augmented from the commencement of the process.

II. But if the enlargement of the vent by the sudden and violent explosions with which the phenomena of an eruption commence, has the effect of temporarily diminishing the repressive force, we cannot but perceive that another important class of circumstances is generated by the action of the volcanic force, tending in the most direct and absolute manner to augment the combined forces of repression, and put a stop to the continuance of the expansive process. We have already observed that such an effect necessarily accompanies that peculiar mode of volcanic expansion (subterranean eruption) which consists in the occupation of fissures, whether *contemporaneous* or *injected*, by intumescent lava; where the weight of the column of liquid occupying the fissure, its refrigeration by contact with the cooler sides, and the diminution of focal temperature, ultimately check any further expansion, and produce the resolidification of the injected mass: by which the repressive forces are repaired and strengthened.

A few considerations on the circumstances that characterise a volcanic eruption taking place on the outer surface of the globe, will convince us that this "suicidal" character, (if I may be allowed the expression,) is still more remarkably exemplified in these cases; and will lead us to the establishment of a most important and interesting general law; viz. that

The development of the volcanic action necessarily and universally tends to its own extinction, by augmenting the opposite forces of repression.

This law will, perhaps, be found, hereafter, to afford an explanation of many of the changes which have taken place in the earth's crust, and which are not at present generally supposed to have any connection with volcanic action.

The circumstances to which this effect is owing, are principally the following, viz.

I. The liquid lava forced up the vent by its own intumescence and that which rapidly takes place below it, increases the length of the vertical column pressing upon the inferior mass, by the whole depth to which it rises within the fissure of escape, and consequently augments the compression upon every molecule towards which the expansion is propagated inwardly, by the weight of a column of liquid lava of that height. The effect of this repressive circumstance will vary directly with the specific gravity of the lava, and the vertical depth of the fissure broken through the solid overlying crust.

II. The surface of the liquid lava propelled up the vent into contact with the atmosphere, parts rapidly with the elastic fluids that produce its liquidity: and in proportion as it thus loses the mobility of its parts and becomes solid, this superficial crust must oppose by its cohesion, a considerable resistance to the expansion and escape of fresh matter from beneath it, over and above that occasioned by its weight.

III. The fragmentary matters ejected by the violent expansion and rise of the elastic fluids that escape from the vent, partly accumulate around its mouth, but in great part fall or roll into it again, and thereby obstruct the rise and expansion of fresh matters. The proportion of those fragments that fall or roll again into the mouth of the orifice after their projection will increase gradually with the sum of the ejected matters, that is with the continuance of the eruption; since the greater the accumulation of fragmentary substances heaped up around the mouth of the vent, the less easily will the fresh fragments ejected be enabled to arrange themselves in a stable manner on the outside of the vent. In fact, the subconical crater that is thus formed around the opening, acts like the mouth of a funnel in directing the falling fragments into the vent; the more the dimensions of this funnel are increased by the addition of fresh materials to it, the greater evidently will be the proportion of fragments directed into the vent.

From hence it inevitably happens that the sum of the forces of repression, which, at first, perhaps even progressively diminished by the widening of the vent, will, sooner or later, by the combined influence of these circumstances, begin again to increase; the ratio of internal expansion, which was at first accelerated, will then be doubly retarded; the discharge of the dilated matters will by degrees diminish, with the internal intumescence that occasions it, and finally cease, when the increase of compression, and the decrease of expansion, have brought these opposite forces to an equilibrium.

The moment at which the compound force of repression begins to increase, forms the *crisis* of the eruption; the violence of which must progressively diminish, as this force increases in an accelerated ratio, while the energy of the opposite force of expansion, which then was at its maximum, is proportionately diminished.

§. 23. We have seen that while the expansion of the lava progresses inwardly, the excess of the intumescent matter must rise up the fissure of expansion; and if this fissure communicate with either of the fluid media, air or water, that outwardly envelope the solid crust of the globe, it will escape absolutely upon the outer surface of the earth.

This escape may take place, according to circumstances, either,

I. By the rise and absolute expulsion from the mouth of the vent of the liquid lava alone—which consists, as we have seen, of an assemblage of crystals of one or more minerals, more or less disintegrated and comminuted, merged in a vehicle of steam, proceeding from the volatilization of the water which was intimately united with their texture. Or,

II. By the partial elevation, and perhaps expulsion, of the lava, and the absolute escape, by itself, of a certain proportion of the elastic vapours contained in it; which traverse the mass of liquid lava, and rise from its surface in consequence of their inferior specific gravity.

The degree in which this escape will be effected, that is, the proportion of the elastic fluids generated which are propelled upwards and escape from the lava, to that which remains confined within it, and occasions the augmentation of its volume, is, as we have seen above, determined by *the liquidity* and specific gravity of the lava—in other words, *by its fluidity*.

§. 24. It is clear that as the dilatation progresses towards the interior of the subterranean mass of solid crystalline lava, the gaseous fluids cannot escape from any depth below the surface as quickly as they are formed, however great may be the fluidity of the lava; which must therefore in all cases intumesce more or less, and rise en masse up the vent.*

---

* It is this law that determines the intumescence of all liquids by ebullition. The degree of intumescence varies inversely with the fluidity—that is, with the specific gravity and *mobility* of the liquid. In the same manner as, in our kitchens, the greater the *consistence* and *lightness* of the substance, the greater the quantity that *boils over*; so, in volcanic eruptions, the degree of intumescence, and consequently the quantity of lava elevated en masse by the same volcanic force, is inversely proportionate to its liquidity and specific gravity.

There exists also sufficient reasons for believing that the lava itself is in all cases elevated, either to the mouth of the vent, or, at least, to within so short a distance from it, that its surface is in free communication with the external air.

In fact, the liquidity of the dilated lava at any point (ceteris paribus) varies universally with the compression it sustains; and, since the pressure or resistance experienced from above by the surface of the column rising within the vent, is one of the elements of the compressive force, the liquidity of the lava, as it rises, must vary inversely with this resistance. But the expansive force of the elastic fluids struggling to rise and escape through the fissure, must, by its reaction from the sides of this aperture, increase the superficial pressure on the lava below, and consequently diminish its liquidity, thereby producing an impediment to the freedom of their own motion, and consequently augmenting the volume of the lava, and forcing it to rise higher, en masse, up the vent. This reaction, and the effect it produces, will be proportionate to the depth, narrowness, and irregularity, of the fissure, and consequently will be diminished by its forcible enlargement. The fall and accumulation of the fragments broken off from the sides of the fissure has a similar influence in diminishing the liquidity of the lava, and raising the level of its surface within the vent—and both these causes united, will, in all cases, necessarily occasion the elevation of the lava to within a very short distance of the mouth of the orifice, and preserve it at that elevation, until the expansive force ceases entirely to be developed below, or is so far diminished that the elastic fluids escape from its surface, by reason of their inferior specific gravity, in a quicker ratio than they are generated below by the retarded progress of dilatation. The level of the liquid lava will then begin to sink within the vent, and subside, till all escape of the confined matters, even in a gaseous form is impeded, by the successive augmentation of the forces of repression.

§. 25. The phenomena rendered sensible on the outer surface of the earth, by this development of the expansive force of lava confined beneath it at an intense heat, constitute a *volcanic eruption.*

This must commence with a series of jarring shocks, or earthquakes, repeated at more or less distant intervals; the last being immediately followed by the more obvious display of subterranean activity at the upper extremity of the rent that has been at length forced completely through the earth's crust; viz. by the violent expulsion and escape from this aperture; either

I. Of a mass of liquefied lava alone, when its fluidity is so extremely *imperfect*, that none of the contained elastic fluids are enabled to traverse it, and escape from its surface, by the influence of their inferior specific gravity. Or,

II. Of the aeriform fluids alone, when the fluidity of the lava is so *great*, that, after its surface has risen to the mouth of the vent, and thus has acquired a free communication with the outer air, its intumescence is carried no farther, but the elastic fluids traversing it with facility, escape from the surface exactly in proportion to the excess of dilatation produced in the lower strata of the liquid mass. Or,

III. Of both the lava and elastic fluids in the proportions determined by the degree of its fluidity.

The first of these three cases appears almost entirely confined to subaqueous eruptions; where the superior density of the ambient fluid, and the greater external pressure it produces, may be supposed occasionally to diminish the fluidity of the lava to this excessive degree.

Again, the extreme fluidity necessary to produce the second case is of rare occurrence, if not solely confined to such volcanos as are in permanent eruption. (Stromboli; and some of the volcanos of the Andes, whose detonations are occasionally heard, but which exhibit no other phenomena.)

In the generality of instances the fluidity of the lava appears to have oscillated between these two extremes, and consequently to have occasioned the absolute escape from the orifice of eruption both of lava en masse, and of the steam evolved from it, in varying proportions determined by the degree of its fluidity.

§. 26. In such a case the external phenomena of the eruption must begin by the violent escape or discharge of elastic fluids from the vent; since the gradual intumescence of the lava up this channel, must occupy more or less time, during which the gaseous fluids are continually escaping from its surface, and rising rapidly up the aperture.

The ratio of the intumescence of the lava, and consequently of its expulsion from the orifice, after it has attained that level, varies, we have seen, inversely with its fluidity, the *expansive force* being fixed. But since the development of the expansive force, or "*eruptive force*," usually increases in energy at first, owing to the enlargement of the vent, the violence of the eruption, and therefore the ratio of expulsion, both of the lava and aeriform fluids, will at first be gradually augmented. The eruptive force, however, diminishing again in intensity, as the repressive forces are in turn augmented, the eruption will shortly reach its crisis; that is, the moment when the eruptive force is at its maximum of violence; after this, the ratio of production will gradually decrease; the lava will soon cease to be expelled en masse from the vent; the contained elastic fluids will escape from it by the impulse of its greater specific gravity faster than they are generated below by the gradually retarded progress of dilatation inwardly.

By this escape, the volume of the lava is diminished, and its surface subsides within the vent. Thus the violent discharge from the volcanic orifice of the aeriform fluids alone, constitutes the concluding phenomena, as well as the commencement of the eruption.

But at the same time that the level of the lava is slowly lowered within the volcanic chimney, the falling of the scoriæ projected upwards by the gaseous explosions, tends more and more powerfully to obstruct this issue. The expansive force of these fluids struggling to escape from the lava when at some depth within the vent, may again break up, and further enlarge its dimensions, and thus momentarily revive the decaying energy of the expansive force. More than one oscillation of this sort will perhaps occur, in which the eruption will appear to be renewed; but as, the more violent the struggle, and the longer it continues, the greater the quantity of obstructing fragments produced,—these brief paroxysms only exhaust the energy of the expansive force, and accelerate the victory of its antagonist.

Many of the fragments ejected in this manner are more or less triturated and pulverized, by the immense friction they undergo in their repeated projection and fall. A great proportion of the finest and lightest are entirely discharged, rising with the escaping gasses into the higher regions of the atmosphere in turbid clouds; but as the surface of the lava column sinks lower, its explosions lose at length the power of projecting even these loose and fine fragments beyond the lips of the eruptive chasm or crater. They must therefore continually accumulate in increasing quantities, both by direct fall, and by rolling down the inner slopes of the crater, upon the surface of the lava, and eventually must completely stifle its ebullition.

The eruption then has terminated. The repressive forces have completely put a stop to the outward development of the expansive force; and they continue to increase for some time by the rapid dilapidation of the deep and ruinous crater, whose debris augments the pressure on the subjacent lava.

The rain-water that collects at the bottom of the crater, and penetrates through the fragmentary matters accumulated there, likewise increases the forces of repression by refrigerating and consolidating the surface of the residuum of lava.

The escape of the steam from its superficial crust by percolation through the fragmentary accumulations that cover it, propagates the consolidation still further.

§. 27. If we now examine what must be the condition of the subterranean body of lava from which the eruption broke forth, we shall find that a definite space of this solid, but heated, mass, has, by the process above described, been more or less disintegrated and dilated; a certain quantity of its crystalline matter, and a still greater proportion of the aqueous vapour generated from the water it contained, having escaped upon the surface of the earth, leaving the remainder in a dilated state from the quantity of elastic fluids still disseminated through it, *and at a greatly diminished temperature.* The circumstances that determine the absolute and relative dimensions of this space, which may be properly called the *focus* of eruption, have been already laid down, (See p. 39.)

The complete obstruction of the vent has now put a stop to the progress of dilatation towards the interior of the mass, which continued only so long as any part of the dilated matters was expelled from this avenue of escape; and the forces of expansion and repression are momentarily in equilibrio on every point of the focus.

§. 28. The temperature of this dilated mass is, however, very inferior to that of the solid lava which surrounds it; consequently the latter will gradually part with its caloric to the former, till an equilibrium is attained. But this gradual augmentation of temperature in the focal mass must proportionately increase the expansive force of the elastic fluids it contains.

The ratio of its augmentation depends on that in which fresh caloric is communicated from the surrounding surfaces, and may be supposed to vary with the temperature of the stratum of crystalline rock of which it forms a part—the conducting powers of the solid lava—the depth of the focal mass, &c.

§. 29. If the ratio of this augmentation should be sufficiently rapid to counterbalance the continual increase of the repressive forces by the greater or less obstruction of the vent, (which under a combination of peculiarly favourable circumstances influencing the disposition of these obstructing matters, we shall presently show to be possible, and occasionally though rarely to take place,) the eruption becomes permanent, the increments of caloric communicated to the dilated mass from its enclosing walls, pass off by the outward escape of gaseous and solid matters, in the exact ratio of their communication; the temperature of the focal mass remaining constant; and the only change to which it is liable being, that since a certain proportion of the solid matter of this lava (whether in a fragmentary or liquid state) must be always expelled from the vent together with the escaping vapours, the consequent diminution of pressure will propagate the expansion proportionably further in depth, by which means a portion of the solid walls surrounding the dilated mass will continually enter into ebullition, exactly corresponding in bulk to the portion which this mass loses by the ceaseless eruptions of the vent. This liquefaction of fresh matter apparently supplies the continual expense of water, which takes place in the volumes of vapour that rise successively from the boundaries of the dilated mass to the exposed surface, and are visibly discharged from the volcanic orifice.

In such a case it is obvious that the eruptive phenomena must be comparatively tranquil and uniform; affording a measure of the ratio in which fresh caloric is constantly communicated to the subterranean mass of liquefied lava, or focus, from the surrounding and solid substance.

This condition of a volcanic vent represents what has been named above, the phase of permanent eruption.

§. 30. But when, as almost invariably happens, the ratio of increase of the repressive force during an eruption, exceeds that in which the expansive force of the dilated mass is augmented by the slow transmission of caloric from the enclosing rock, the moment must arrive when the eruption terminates. The fissure of eruption is finally obstructed, and all absolute escape of the dilated matters prevented.

In this state of things, the continual increase of the expansive force of the liquefied mass must have the effect of narrowing its limits—the outer and lower parts, in which the increase of temperature is greatest, gaining in volume at the expense of the inner and upper parts, where the increase of temperature is slowest, and which must be proportionately compressed, and ultimately solidified by the condensation of their enclosed vapours. By this process the former vent or fissure of escape becomes still more durably and efficiently obstructed; it is as it were hermetically sealed by the consolidation of the lava which occupies it.

§. 31. But at the same time the lower parts of the focal mass are suffering a gradual increase of their expansive force, and this increase is limited only by their reaching an equal temperature to that of the surrounding mass.

Before this can be effected the increased expansive force of the focal matter will probably overcome the resistance opposed by the overlying obstructions, produced by the last eruption, and by creating one or more ruptures across them, renew the eruption from the same or nearly the same point on the earth's surface. The repetition of such eruptions from the same vent or fissure originally broken through the pre-existing superficial strata, occasions what is called an *habitually active volcano*, and at length by the accumulation of their products gives birth to a *volcanic mountain.*

§. 32. If the increasing expansive force of the focus does not succeed in forcing a fresh issue, before its temperature has been equalized with that of the enclosing crystalline rock, it will then have become completely solidified, (with the exception perhaps of some few points on which the vapour it contained has been forced to concentrate itself,) and assimilated to the remainder of the mass, except in the size of its component crystals, which have been more or less disintegrated.

§. 33. When thus resolidified and equalized in tension to the enclosing mass, the focus shares equally in the general expansive force of the whole stratum of heated rock in which it was formed. The continual increase of this force must sooner or later, either cause the dilatation of the mass, and, by so doing, elevate all the rocks above, or burst a passage for the outward escape of its concentrated caloric. This will take place of course on the weakest point of the overlying rocks, which, in the generality of cases, will be coincident with that where the prior eruption found a vent; and hence arises the far greater frequency of habitual volcanos than of single eruptions from fresh apertures.

§. 34. Sometimes, however, it may happen that the solidification of the lava formerly injected into the eruptive fissure has sealed it so effectually that the subterranean expansive force finds less resistance on some previously unbroken point of the overlying strata; and an absolutely new vent of eruption is thus produced on the globe's surface. This will usually be effected on the prolongation of the original fissure, which must be the direction in which the rocks will yield most easily, having probably been shattered along that line by the earlier shocks. This law gives rise to the long linear trains of volcanic vents so often noticed.

§. 35. The distance of the new vent from the former one will be of course dependent on local circumstances in the structure, tenacity, and other elements of resistance, of the overlying rocks. It may vary from a few yards to many miles. The expansive force, by which these ruptures are produced, being general to the whole underlying bed of intensely heated crystalline rock, the horizontal extent of which is by our hypothesis unlimited; and of which the expansive power is only limited by the local circumstances that upon different points may diminish its tension.

These circumstances will be considered presently.

§. 36. On the whole it appears that wherever an habitual volcano exists a constant struggle is kept up between the expansive force of the subterranean bed of lava, and the repressive force created by the solid and fragmentary matters which accumulate within and over the vent.

Eruptions break out and cease; recommence, and are again stopped alternately, at intervals of greater or less duration, according as the one or the other of these opposing forces acquires the preponderance: the exercise of that superiority in each tending to augment the power of its antagonist. Every volcanic eruption, in fact, is *the means by which the caloric emanating from the interior of the globe passes off into outer space* through the surrounding atmosphere, each volcanic vent acting as a safety valve to the globe. But this outward development of internal energy is necessarily productive of circumstances by which its continuation is in turn impeded, and the temporary stop thus generally put to the transmission of caloric, concentrates it below the impeding surfaces, and augments the energy of its expansive force beneath them, till it again forces a passage.

When the two opposing forces of expansion and repression oscillate for any length of time about an equilibrium, each alternately acquiring the superiority at brief intervals, the volcano exists habitually in the second phase, or that of brief intermittences.

When on the contrary circumstances favour the excess of the forces of repression, the intervals of rest are of long duration, and the violence of the eruptions that break forth when the force of expansion has at length attained the superiority, will be proportionately great, in consequence of the intense temperature to which the confined lava had been raised in the mean time, even in its superior strata.

Such an eruption acquires the character of a paroxysm, and the volcano in which these circumstances are habitual, exists in the third phase, or that of prolonged intermittences.

§. 37. But to produce an extremely violent or paroxysmal eruption it is not necessary that the volcanic vent should have been wholly inactive for a considerable time. On the contrary, it may have existed previously either in permanent or frequent activity, in either the first or second of the phases noticed above. The minor eruptions of these phases, with all their accompanying phenomena, having their origin in upper or secondary foci, in all probability above the level of the vent originally broken through the superficial strata, and within the chimney of the volcanic mountain.

If these eruptions do not carry off the whole excess of caloric communicated to the inferior bed of crystalline and heated lava, the *general* expansive force of this mass, being continually on the increase must eventually overcome the opposing resistances, and break out in earthquakes, and heavings of the whole overlying mass. When by these a fissure is formed penetrating to a great depth within the lava rock, its intense temperature will occasion a proportionately violent expansion, and produce a paroxysmal eruption.

Hence it is that many habitual volcanos appear to exist at once in the second and third phases, and to have a double system of operations, productive both of paroxysmal and minor eruptions.

The first occur at intervals of very long duration, proceed from an inferior focus, take place with intense violence and rapidity, and leave a prodigiously wide and deep crater.

The second class of eruptions, proceeding from higher and secondary foci, throw up cones in the centre of these gigantic craters, which they tend gradually to fill up and obliterate, and continne in very moderate but frequent activity, until their products are once more blown into the air, by an explosive eruption of the first class.

These alternate developments of activity, from the primary or inferior, and the secondary or upper seats of expansion, explain the so frequent appearance of a recent volcanic cone rising out of a larger and older crater, as Vesuvius, from that of Somma;* the Peak of Teneriffe and Chahorra, from the old circus described by M. de Buch;† the different successive circus's of Bourbon;‡ that of Volcano;‖ of the Lake of Ronciglione, of Astroni; that of the N.W. extremity of San Miguel, which is 15 miles in circumference, and contains two or three parasitical cones on its interior;§ Tunguragua, at the bottom of which Humboldt saw some minor cones in full activity, &c. &c.

§. 38. The nature of the circumstances which determine the relative power of the two opposite forces of expansion and repression, and consequently, the energy and duration of eruptions, presents an interesting subject of investigation, which it will be as well to pursue somewhat more in detail.

By the hypothesis with which we set out, the accession of caloric from the interior of the globe towards any individual bed of lava in the vicinity of its surface, was supposed to be constant and uniform.

Therefore the conduct of the expansive force would of itself be regular and invariable. And we must look for the origin of the irregularities in the phenomena in question to variations of the repressive force alone.

And first for those remarkable instances of external volcanic action, in which the opposite forces are so nicely and continually balanced, that the ebullition is permanent, and the volcano in the first phase.

Stromboli has been already cited as a rare (and perhaps almost unique) instance of ceaseless eruption. But it is not

---

* † ‡ ‖—See the succeeding Figures.

§ *Webster's Description of St. Michael, Boston.*—1822.

difficult to discover the peculiar circumstances to which this volcano owes the remarkable constancy of equilibrium that is here preserved between the expansive force of its focus, and the forces of repression mentioned above. Dolomieu and Spallanzani both mistook the character of this volcanic mountain. They supposed the vent to have formerly existed at some distance from the actual one, because the latter is overlooked by a much higher eminence, and these Geologists had been accustomed to find the chief crater of a volcano uniformly at its summit. The fact however is, that the orifice of eruption lies as usual at the bottom of the crater, and has never changed its position in a lateral direction; but one whole side of the crater has evidently been destroyed at a distant period; probably blown up into the air by some violent paroxysm, which may be supposed to have preceded the present phase of moderate and prolonged activity.

The present cone of Stromboli has in consequence a remarkable figure, which will be best understood by reference to the accompanying profile, section, and plan.

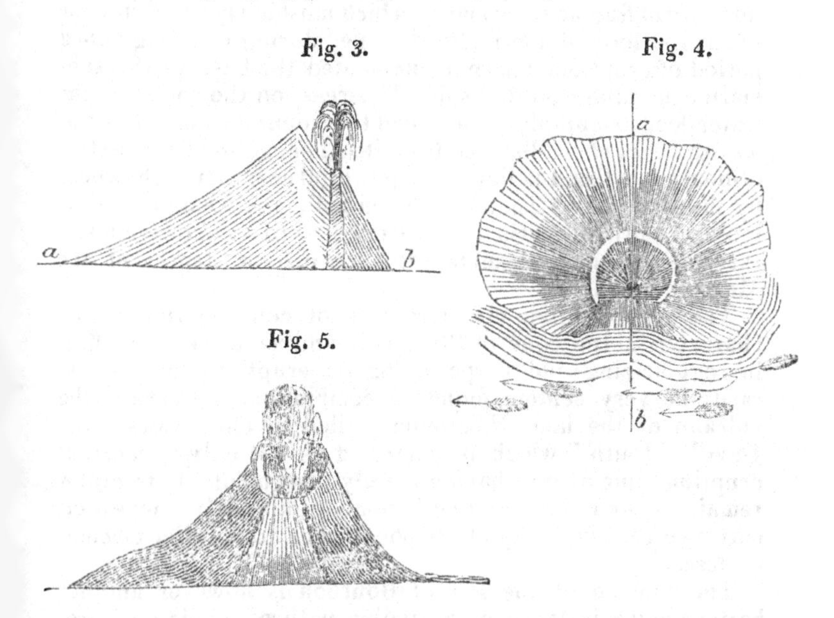

Fig. 3. Fig. 4. Fig. 5.

The orifice of eruption is bounded on one side by a high semicircular slope, forming the remaining part of the crater, while immediately from its lip on the other side, a steep slope commences, reaching to the sea, and inclined at an angle of 45°; consequently too rapid for any fragments but the lightest and smallest to stop upon it.

Hence, of the fragments thrown up by each explosion from the vent, less than one-half can be supposed to be returned into the orifice, either by direct fall, or by rolling down the sides of the semi-crater; while the remainder falling on the other side of the vent, roll with excessive rapidity down the inclined plane into the sea. It is clear therefore that owing to this peculiar conformation, a very large proportion of the fragments created by the bursting of the steam bubbles through the surface of the liquid lava are irrevocably discharged from the vent at each explosion, and the surface of the lava within can seldom be covered but by a limited quantity at a time, which quantity cannot be increased by the continuance of the eruption.

Since it has been said that there is reason for supposing Stromboli to have existed in its present phase at least for the last 2000 years, it may be thought strange that the immense quantity of fragmentary matter which must have rolled into the sea at the foot of the inclined plane, during this lengthened period of eruption, has not augmented the base of the submarine mountain on that side. Whereas, on the contrary, the water deepens rapidly at the foot of the mountain, and is fathomless at a very short distance from it. The cause of this extraordinary fact lies, I believe, in a powerful current which sweeps by the side of the island of Stromboli, and wears down the loose and brittle scoriæ that are exposed to its abrasive force, carrying off the fragments into the distant depths of the Mediterranean.

The very peculiar circumstances of conformation which characterise the cone of Stromboli, and have such an effect in regularizing and perpetuating its eruptions, will necessarily be very seldom found in combination. Perhaps the volcano of the lake Nicaragua, called by our sailors "the Devil's Mouth," which is described as not only in constant eruption, but also as having merely a segment of its crater remaining, and as rising, like Stromboli, abruptly from water, may owe the similarity of its phenomena to parallel circumstances.

The volcano of the Isle of Bourbon is however another better known instance of a similar nature. This vent frequently remains for the space of many weeks in ceaseless eruption.

In this instance the form of the cone is perfect, and we

must look for the cause of the permanent superiority of the expansive force in other circumstances; nor is this a difficult task. The appearance and phenomena of this volcano have been described in considerable detail by M. Bory de St. Vincent; and from his drawings and observations it appears that no central crater or cavity of any size exists upon its summit (see fig. 7); but three or four apertures are occasionally formed, and during the continuance of the eruption are filled again. He gives one particularly interesting sketch of the "Cratere Dolomieu," as seen by him in the night time (see fig. 6), from which it appears, that during continu-

Fig. 6.

Fig. 7.

ous eruptions the ebullient lava remains at a level with the summit of the cone, and is continually boiling over the lips of the vent, like any thick or viscous liquid over the edges of a circular vessel in our kitchens, spreading on all sides in concentric undulations from the centre, where it is seen to rise and fall alternately in a liquid state, and at an intensely white heat, owing to the continual rise of large bubbles of vapour which escape in succession from its interior. The scoriæ, or tattered lava fragments, thrown up by these aeriform explosions, are in much less proportion than at Stromboli, and consist of filamentous pomice, frequently drawn out to the finest threads, like spun glass (verre capillaire volcanique); instead of those rude and heavy fragments of augitic basalt which are ejected by the other volcano.

§. 39. It is highly important to remark the difference between the phenomena of these distant volcanic vents during their permanent state of eruption; for, notwithstanding these discrepancies, it will plainly appear that they are regulated by the same laws; viz. those which have been laid down above, from which these discrepancies will be seen to flow of necessity.

In Stromboli, the gaseous explosions are ceaseless and violent, while the lava never boils over the lips of the orifice.

In the Isle of Bourbon the lava is almost continually in extravasation over the edges of the vent, and the discharge of aeriform fluid comparatively trifling.

The lavas of Stromboli have a high degree of liquidity, as is manifested by their cellular nature, and an extremely high specific gravity, being solely augitic—their fluidity therefore is proportionately high.

The lava of Bourbon, on the contrary, with a great degree of liquidity (its disintegration having proceeded almost or entirely to the integrant molecule, since the lava appears in complete fusion, and hardens into a glass), possesses a very low *specific gravity*, since it consists solely of felspar. Hence its fluidity must be proportionately low.

But we have observed above, that the proportion of vapour which is propelled upwards, and escapes from any lava, to that which remains behind, varies with its fluidity. It is therefore evident why the feldspathose lava of Bourbon boils over the lips of its vent, while the heavier lava of Stromboli allows of the rise and escape of all the vapour from it as fast as it is generated, and preserves the same level at the mouth of the orifice, without any extravasation.

The glassy lava that thus is continually poured over the edges of the orifice on the summit of the volcanic cone of Bourbon, forms numerous small streamlets, seldom more than a foot wide, that descend the sides of the cone. They are frequently hollow, like a pipe or arched gutter, the interior of the stream having continued to flow on after the supply had ceased, and the external crust had been consolidated. This thin crust is often broken through by the foot.

The Mamelon Centrale, a small parasitical cone, which has been elevated in the vicinity of the Cratere Dolomieu, by previous repetitions of the same process described above, illustrates these observations in a beautiful manner.

Fig. 8.

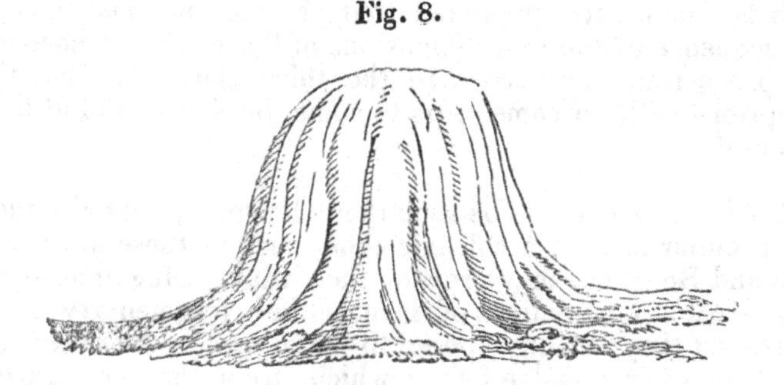

But this constant extravasation of the lava over the edges of the orifice of production, is obviously the cause of the permanence of the eruption; since, by the escape of the elevated lava in the exact ratio of the intumescence that takes place below, the repressive forces are preserved in equilibrio with that of expansion; while at Stromboli the same effect is produced by the continual removal of the due proportion of elevated matter, only in a fragmentary instead of a liquid form.

§. 40. These examples suggest the consideration that the oscillations about an equilibrium between the forces of expansion and repression will be more or less marked, producing intermittences of longer or shorter duration, in proportion as the circumstances predominate which favor or impede the complete escape of the protruded matters from the vent.

It follows from this, that the greater the dimensions of a volcanic crater, the longer should be its intervals of repose;

and we should expect that as the crater becomes gradually filled by the quantity of matter accumulating within it, the frequency of the eruptions will increase.

The truth of this proposition was strongly exemplified in the numerous and almost permanent eruptions that took place from Vesuvius between the years 1804 and 1822, when the old crater of 1794 was nearly filled; and which increased in frequency as the summit of the cone, in place of a concavity, acquired by degrees a convexity of surface; which might have preserved the vent in ceaseless eruption, like that of Bourbon, but for the violent paroxysm of 1822, which entirely changed the figure of the cone, and excavated an enormous chasm through it. If these considerations are just, it may be confidently predicted that the volcano has now, in consequence of the vast dimensions of this cavity, exchanged its phase from the second to the third class, and that the eruptions will, for some years to come, be fewer, and at long intervals.

§. 41. It remains to be remarked, however, that although the peculiar form of a volcanic cone, such as those of Stromboli and Bourbon, may prevent their vents being choked by any great accumulation of lava, either fragmentary or in mass, yet they remain exposed to variations in the other elements of the repressive force; which, from the comparative tranquillity and uniformity of their phenomena, will be more obvious in these volcanos of permanent action, than in others of which the causes are more complicated. Such a variation will, it is clear, be necessarily produced by changes in the pressure of the atmosphere; and, however bold or novel an assertion it may appear, to suppose the subterranean action of a volcanic vent to follow all the minute variations in the weight of the atmosphere, yet, on consideration, it will, I think, appear that such must in reality be the case; and I have little hesitation in expressing an opinion, that the phenomena of a volcano in permanent, or nearly permanent, eruption, are modified by all the changes of atmospheric pressure, as effectually and constantly as the mercury in a barometer—not only that the lava rises or falls in the vent, as the mercury in the reservoir of the barometer, but that the boiling point of the water contained within the lava must vary with the weight of the atmosphere, and its ebullition be checked by an increase of weight, and augmented by a diminution, as truly as that of the water contained in the ingenious barometer for measuring heights, of Dr. Wollaston.

In fact, the weight of the atmosphere is one essential element of the repressive force by which the extent of the dilated mass of lava is determined; any diminution, however trifling, in this weight, must allow the process of ebullition to be propagated some way farther, and proportionately augment the intensity of its action; and, again, when the atmosphere returns to its previous density, the slight augmentation of pressure on the surface of the column of liquid lava, will check the ebullition of the lower strata of the dilated mass, until their temperature has again risen to the boiling point corresponding to the increased pressure.

If from theory we turn to observation, we shall find that it has been a frequent subject of remark, that the development of volcanic energy is often in a remarkable degree connected with the condition of the atmosphere.

The inhabitants of Stromboli positively make use of the volcano as a weather-glass. They are mostly fishermen; and while engaged in their occupations at a short distance from the island, have its orifice constantly in view; and I was assured, by all whom I questioned on the subject, that its phenomena decidedly participate in the atmospheric changes, increasing in turbulence as the weather thickens, and returning to a state of comparative tranquillity with the serenity of the sky.

During the tempestuous weather of the winter season, the eruptions of the volcano no longer preserve the uniform march which characterises them the greater part of the year. The explosions are then often so terrible that the island seems to shake from its foundations; and these paroxysms, which sometimes last for days, are succeeded by intervals of complete quiescence, of a few hours duration, which are in turn followed by other eruptions of similar energy. On such occasions the steep flank of the cone, which forms the inclined plane beneath the crater, has been occasionally observed to be rent in two by a vertical fissure, which emits a torrent of lava into the sea. The chasm is afterwards sealed again by the consolidation of the lava, and is covered and concealed by the loose fragments ejected from above. After eruptions of this energy a quantity of fish have often been thrown up by the sea on the beach of the island, apparently killed by the sudden heating of the water, since their flesh is found to be loosened from the bones just as if boiled.

The connection thus observed between the intensity of the volcanic energy in habitually active vents, and the condition of the atmosphere, has been frequently remarked in other instances.*

§. 42. But this connection is not peculiar to the development of the force of subterranean expansion by an open volcanic vent, for the changes of atmospheric pressure appear often to influence in a similar manner those more occult modifications of the same force which are exhibited in earthquakes. These have been frequently found to take place in stormy weather, during hurricanes, and particularly in the winter months.

If it appear probable and obvious that a diminution of atmospheric pressure on the narrow mouth of an active volcano, will have a sensible effect in augmenting the ebullition of the confined lava, it is surely not incredible that a similar change acting contemporaneously upon a vast extent of the globe's surface, by diminishing the sum of the resistances that counteract the powerful expansive force of a bed of lava confined at an immense depth below this surface, must determine the partial development of that force, by which these shocks are, as I conceive, produced.

In fact, the subterranean force of expansion, ever active, and continually pressing upwards with a gradually increasing energy, must be frequently restrained by only the slightest degree of superiority in the forces of repression, so that the least imaginable diminution in the elements of the latter force may occasionally suffice to give the predominance to its antagonist.

It is obvious how the powerful ascending draught of air which constitutes a hurricane, and which acts so strongly in depressing the barometer, will have an equal effect in setting loose the imprisoned winds of the earth.

The remarkable coincidence which has been so frequently observed between disturbances of the atmosphere and subterranean commotion of every kind, is thus accounted for by the principles of volcanic action laid down in this essay, with a simplicity and facility which, it is presumed, are strongly conclusive as to their correctness, and the justness of the hypothesis upon which they are founded.

It may as well be noticed in this place, that, while the changes in the condition of the atmosphere have in this manner a very sensible and powerful effect in modifying the

* The Peak of Ternate, in the Moluccas, is said to break out with greatest violence during the Equinoxes.

volcanic phenomena; these, in turn, re-act upon the atmosphere, producing meteoric phenomena of great importance.

They consist in

I. *Hurricanes*, or ascending currents of air, caused by the rapid rise of so large a column of heated vapour.

II. Rain in torrents from the clouds of aqueous vapours that mount up from the vent.

III. Electrical phenomena—probably in great part occasioned by the violent friction of the ejected fragments against one another, as they are driven forcibly upwards with intense velocity.

§. 43. It appears, therefore, that the chief circumstances by which the quiescent intervals of habitually active volcanic vents are produced, consist in the weight and cohesive force of the rocks, both fragmentary and solid, which accumulate above and around the orifice of eruption, and that but for this tendency of every eruption to impede its own continuance by the production of these matters, every volcano on the surface of the globe would exist in permanent and nearly regular eruption; no earthquakes would take place: but the escape of caloric from the interior of the globe into outward space in combination with aqueous vapour, and through the vents of this natnre, which are scattered in great numbers over its surface, would be comparatively constant, tranquil, and invariable.

§. 44. But in conformity with the important law mentioned in a former chapter, the outward development of the subterranean force of expansion is, in almost every instance, temporarily stopped by obstacles which it creates to itself. From hence must necessarily result a great irregularity in the conduct of the caloric of the interior of the globe. Wherever its escape is checked it must accumulate. This accumulation tends to dilate the rocks in which it resides, and by producing fractures in the overlying stratified crust, to afford a fresh avenue of escape for the caloric, under some form or other.

In this manner the draught of caloric is forced often to shift its direction, and is diverted from one vent to another.

The circumstances by which these alterations are produced, come now under our consideration.

§. 45. The liquefied mass (or *focus*) left by any eruption, continually abstracting caloric from the solid lava which encloses it, proportionately retards the augmentation of the expansive force of its inferior strata, which only increase in temperature as long as they are unable to part with their excess of caloric, in the ratio in which they receive it from below.

Fig 9.

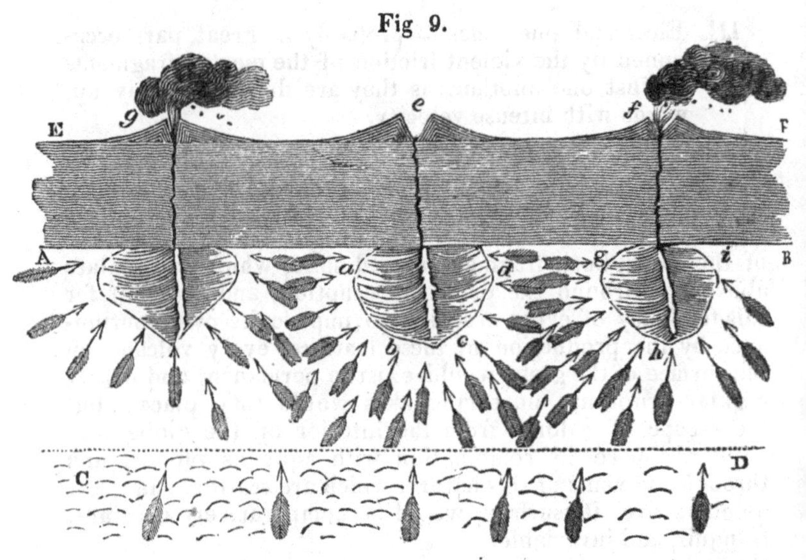

Thus, let *a b c d* be supposed the section of the liquefied lava mass, or focus, left by the eruption which produced the cone *e*, in the subterranean mass of heated lava, A B C D. *a b c d* has been reduced by ebullition to a lower temperature than the enclosing mass, and therefore abstracts caloric from it in the directions indicated by the arrows.—Consequently the inferior strata of this mass, viz. C D, part with their caloric more quickly than before the eruption. If the vent remain permanently open, the draught of caloric thus created is regular, uniform and constant. If the vent is shut by the predominance of the repressive force, the draught of caloric towards the focus *a b c d* takes place in a continually retarded ratio. In the former case, the caloric may make its escape outwardly in the same ratio in which it is communicated from below to the inferior strata of the lava mass, in which consequently there will be no increase of expansive force; or it may escape in a lower ratio, and the expansive force of C D will be continually on the increase, notwithstanding the constant expulsion of a certain proportion of caloric through the permanently active vent.

In the latter case, that is, when the vent *e* is closed, the draught of caloric towards *a b c d* taking place in a gradually retarded ratio, it cannot prevent the constant increase of temperature, and consequently of expansive force in C D, the result of which, under both these last circumstances, must be its successive dilatations, and the progressive elevation both of the overlying solid mass A B, and of the superficial strata E F, producing fractions and dislocations in them, through one or more of which a fresh eruption may, or may not, take place. This will either be on the same point (*e*), or on the continuation of the fissure upon which the vent *e* was formed; or, at any rate, the new fissures will most probably be parallel to the old, in consequence of the overlying strata splitting most readily in the same direction.

If we suppose such a vent of escape to produce another eruption from a fresh point *f*, the same results will be produced as at *e*; but it is obvious that the draught of caloric created by this new focus *g h i*, not only lessens the increase of temperature of the underlying strata C D, but also diminishes the quantity of caloric laterally transmitted to the older focus *a b c d*.

Hence we deduce the proposition, that the creation of a fresh vent, in the vicinity of a pre-existing one, diminishes the ratio of this increase of its expansive force; and generally that of two or more neighbouring vents—each retards the augmentation of the expansive forces of the others, in direct proportion to its activity, that is, to the quantity of caloric which passes off by it in given times, whether by permanent or intermittent eruptions.

It is obvious that by the draught of caloric thus created in two directions, the intermediate space *d g c h* is proportionately cooled down, and by the activity of these neighbouring vents, this refrigeration of the intermediate space may continue so long as to diminish effectively its expansive force, and obviate all elevation or rupture in the overlying rocks.

If we conceive a third eruptive vent *g* to be produced by the subsequent efforts of the expansive force of C D, it is clear that the former focus *a b c d*, being interposed between two active foci, each requiring a continual lateral draught of caloric, will lose proportionately in its supply, and consequently the increase of its expansive force be proportionately lessened. The loss thus sustained may ultimately entirely check the increase of the expansive force of the focus *a b c d* so as to give the preponderance to the forces of repression on that point, and cause the complete *extinction* of the vent *e*.

Should, however, either of the neighbouring vents *f* or *g* be afterwards itself permanently closed by the predominance of the repressive force, the draught of caloric may be again borne towards the focus *a b c d* in sufficient quantity to produce a renewal of its activity.

Thus it is that volcanic vents are either temporarily or completely extinguished, the draught of caloric having been led to take another direction.

Besides the habitual eruptions of such vents, other avenues for the escape of caloric may be afforded in the more placid emanations of heated vapour by percolation, thermal springs, or even in the gradual transmission of caloric to the outer atmosphere or ocean, through the superficial rocks.

§. 46. By these varying modes the caloric communicated from below to the subjacent mass C D, passes outwardly. So long as this transmission takes place in the same ratio as it is received, the expansive force of C D, ("general or primary subterranean expansive force") remains invariable.

When, (as must generally be the case, in consequence of the law we have already determined,) these prove inefficient for the purpose, the subjacent mass C D must increase in temperature and expansive force in proportion as the ratio of the accession of caloric exceeds that of its deperdition.

This continual increase must produce repeated rupture and heavings of the overlying mass, by which the upper parts of the solid lava mass together with the foci or liquefied parts they contain, as well as the hills or accumulations of erupted matter produced by their several vents, and the superficial strata of the globe through which these several vents were formed, will be progressively more or less elevated.

It thus becomes necessary to distinguish between these expansions of the inferior strata of the general mass of heated subterranean rock, brought on by its continual increase of temperature, from receiving more caloric than it can part with through the overlying rocks, or by means of volcanic vents, and the local expansions of minor and less deeply seated foci. Both occasion the same external phenomena;

viz. earthquakes,* the elevation of overlying rocks, the formation of veins, injected and contemporaneous, as well as external eruptions of dilated matter on the surface of the earth, these effects differing only in the energy of their development, which is proportioned to the tension, and consequently to the depth of the expansive force which produces them. The first may be distinguished as the *general*, or primary subterranean expansive force. The second as *local*, or secondary forces.

These secondary expansions, when in sufficiently constant activity, may check and prevent the development of the inferior primary force, by conveying away the caloric as fast as it emanates from the interior of the globe; and in general, the activity of the primary force of general expansion will be inversely proportionate to that of the secondary, in the same part of the globe.

This is a conclusion, to which we shall have occasion to revert hereafter.

§. 47. The laws which have been thus far investigated, and which appear necessarily to determine the apparently irregular and unequal developments of volcanic energy, (under our hypothesis), hold equally good, whatever be the scale of magnitude upon which these phenomena are developed.

The slight and scarcely perceptible tremors which precede any insignificant eruption of an habitually active volcano, are produced by the same effort of subterranean expansion, which, acting on a larger scale, from a focus at a greater depth, and in opposition to infinitely more powerful forces of resistance, occasions those tremendous vibrations of the crust of the earth which extend across a quarter of the globe, spreading ruin and desolation upon their track; and to whose successive developments, the elevation of continents above the level of the ocean, and that of most mountain chains above the average level of the continents, must perhaps be attributed.

---

* "Werner distinguished two classes of Earthquakes;

1. Those which are locally connected with some active volcano, are felt only within a radius of some miles from it, and usually correspond in time with the paroxysms of the mountain.

2. Those which seem to have their focus at a far greater depth, of which the effects are on a much larger scale, and which are propagated with vast celerity to an immense distance, often of some thousand miles."—

Daubuisson Essai de Geognosie, tom. 1.

The same class of circumstances which, in an habitually active volcano, repress the volcanic energy of the subterranean *local focus*, and put a stop to its outward eruption for a few hours, will, when carried to excess, create a quiescent interval of many centuries. In the same manner as a narrow fissure may be obstructed by the congelation of the vein of lava it contains, or by the accumulation of loose fragments above it, and the local expansive force of the lava below checked for a time, and forced at length to break out by another issue: the accumulation of similar obstructions, of greater magnitude, and during a longer period of eruption, may, and without doubt does in reality, check the general expansive force of the subterranean caloric for intervals of immense duration, and forces it at length to find an issue in some other direction, at a superficial distance perhaps of hundreds of miles, the former vent remaining to all appearance for ever extinct.

But in order to expose with greater clearness this universal identity in the laws of volcanic action, we will, in the next chapter examine more at large the disposition of the solid and fragmentary substances produced by any volcano, (*vent of subterranean caloric*,) and we shall then be enabled to see the manner in which their accumulation tends sooner or later to obstruct this issue entirely; and, therefore, since the expansive force cannot be for ever repressed, force it to take another direction.

CHAP. III.

Disposition of Volcanic Products.

§. 1. It has been mentioned that many volcanos appear to have broken out from beneath rocks of all characters, primitive, transition, secondary, and tertiary. We are therefore at liberty to suppose the existence of the volcanic force in full activity beneath any or all these different formations, stratified or not. Let us imagine, that after perhaps many repeated partial dilatations, by which the rocks above are more or less displaced, heaved upwards, and fractured, some fissure is, at length, prolonged entirely through to the outer surface, affording one or more apertures for the escape of the matters confined below, and a volcanic eruption takes place.

It must be remembered that we are at present confining our attention to the case in which the point of eruption is subaerial; no accumulation of water existing above it.

The case of sub-aqueous eruptions will be considered separately.

§. 2. The eruption commences externally with violent explosions of aeriform fluids, that force their way through the aperture, driving upwards in a vertical direction fragments torn from its sides, in greater or less quantity, according to the incoherence or solidity of the rock of which these are composed. These gaseous explosions will appear to increase continually in violence for a certain time, during which the surface of the subjacent lava rises up the fissure; as it approaches the mouth of this vent, the volumes of aeriform fluids bursting from it project upwards considerable masses of liquid lava at a white heat, which are broken into showers by the resistance of the air, and being rapidly congealed as they ascend, both by contact with so much colder a medium, and by the immediate expansion and escape of the vapour to which they owed their liquidity, fall in the form of solid fragments (scoriæ) of various sizes and fantastic forms. These, together with blocks detached from the sides of the vent, accumulate round the edge of the orifice, forming a circular or elliptical ridge; which increases, by the continued projection and fall of fresh matter, into a hillock, usually of considerable magnitude, whose dimensions however will of course be proportioned to the quantity of loose materials ejected, and this will vary directly with the duration of the eruption, and inversely with the fluidity of the lava. (See p. 27.)

The form of this hillock it is obvious will approximate to that of the solid generated by the revolution of an obtuse scalene triangle round a perpendicular axis erected on some point of its base. The inner and outer slopes of this hill will subtend nearly the same angle to the horizon, and this angle will depend principally on the size of the fragments thrown out, and their disposition to cohere, either from the heat they have retained, or their cementation by mineral substances deposited from vapours which may traverse them, rising from below. The hill thus formed around an orifice of eruption is properly called a *Volcanic cone*, from its having usually more or less the figure of a truncated cone. The cavity it circumscribes, and whose form will approach to that of an inverted cone, or funnel, is the *crater*. Its diameter will depend principally on the size of the *vent*, which will be proportionate to the violence of the eruptions; that is, to the explosive energy which is developed in fixed times.

The depth of the crater, when perfect, will be determined by the ratio of the diameter to the quantity of ejected matter; but many causes, as we shall see hereafter, tend to fill up the crater, besides the obvious one consisting in the constant dilapidation of the loose substances composing its sides, and it will therefore seldom retain this regularity of form.

The circular line or *edge* in which the inner and outer slopes of the cone meet, may be with propriety called the *ridge*, or circumference of the crater.

§. 3. The form of a volcanic cone which, under the most favorable circumstances, approaches, as we have seen, to a truncated cone containing a funnel-shaped concavity, will be almost always variously modified by one or more of the following disturbing causes:

1. The vent of eruption produced by the sudden disruption of the solid rocks overlying the volcanic focus will, in almost every case, have more or less the form of a fis-

sure, that is, have one of its horizontal dimensions much greater than the other; and though the volumes of elastic fluids evolved from below, tend to assume a spherical form in their escape, yet they must be forced to conform in a degree to the shape of the aperture through which they struggle upwards; while, in their turn, by widening the vent, they tend to reduce its figure to the cylindrical. The necessary result of this mutual interference is, that the cone of scoriæ thrown up around the volcanic opening will be somewhat elliptical; its longest axis coinciding with the direction of the longest dimension of the vent. When the length of the fissure is considerable in comparison to its width, the gaseous explosions will escape upon more than *one point*; and tend to produce as many cones: which will really be the result when these points, or *vents*, are so far removed from each other that their spheres of projection do not interfere.

When they are on the contrary so near that their spheres of projection *cut* one another, the hillocks produced will be more or less irregular, in proportion to the degree in which the ejections of each vent are mingled. The extreme proximity of many vents tends to produce a long ridgy hill, in which all resemblance to a regular cone is lost.*

* A fine field exists for the study of the varieties of figure assumed by the hillocks of scoria thrown up by single eruptions, and of the circumstances to which they owe their numerous modifications, in the ci-devant French provinces of Auvergne, the Velay, and the Vivarais. The chaine des Puys, near Clermont, (Dept. Puy de Dome,) offers above 60 volcanic cones strung together on nearly the same line, and reaching about 12 miles in length; and again, in the continuation of the same chain through the provinces of the Velay and Vivarais, upwards of 200 similar cones are closely scattered upon a narrow band scarcely 20 miles in length. From amongst this great number, many examples may be observed of *compound* cones, evidently thrown up from two, three, four or many more points of explosion on the same fissure. That these eruptions were simultaneous, is in many instances attested by the circumstance that the matter projected from each of any two proximate vents had an equal effect in disturbing the regular disposition of those thrown up by the other. When, on the contrary, eruptions from proximate vents have been consecutive, the cone produced by the last eruption will appear to have been superadded to that thrown up by the first, and will itself remain entire and undisturbed.

And examples of this occurrence are equally frequent.

In the Chaine des Puys, the Puy de Montchar, the groupe of the Puy de la Rodde, and that of the Puys Noir, de la Vache, and de las Solas, may be instanced as compound hills, each thrown up, contemporaneously, by

2. A second disturbing cause exists in the accidental unevenness of the ground surrounding the vent, upon which the ejected fragments fall and accumulate. Whatever slope or irregularity of surface occurs within this space, the form of the cone raised upon it must of course be proportionately modified by that of the substratum.

3. Violent winds prevailing in any one direction during the eruption will cause the ejected fragments to fall most plentifully to leeward of the vent. According to M. Moreau de Jonnès the fragmentary matters thrown up by the volcanos of the leeward isles have accumulated

---

three different, but neighbouring vents.* The row of heights extending northwards from the base of the Puy de Louchadiere affords an example of a ridgy string of cones produced from vents so close as to confound their ejections, but still sufficiently removed for each cone to retain a certain degree of regularity towards its centre, by which its individuality can be recognised. The Puy de Laschamp is a long narrow-backed ridge (such as the Italians appropriately call schièna d' asino) evidently thrown up by the simultaneous action of numerous vents on the same fissure so close to each other that their products are completely confounded. Traces remain on this hill of two craters only, which were consequently the latest in eruption.

It is remarkable that the long axis of this ridge and of the other groups mentioned above, as well as those of the neighbouring isolated cones, most of which are more or less elliptical, have universally the same direction with the general chain of which they form a part; and which, as we shall see hereafter, probably attest the direction of the original or *primary* fissure, in which the overlying rocks first yielded to the general and deepest force of expansion. Parallel examples might be quoted in great numbers from the chain of the Velay, but that, besides being unnecessary, few of the hills of this district are distinguished by name; and they both lie less in the way of examination than those of Auvergne, and are also less recent and fresh, appearing to have lost more of their original form by subsequent denudation. The sides of Vesuvius afford an instance of the production of five small cones on the same fissure immediately above Torre del Greco. They were created successively by the eruption which destroyed a part of that town in 1794. In fact all the parasitical cones thrown up by lateral eruptions on the flank of a volcanic mountain are of this simple character, and usually of an elliptical form.

In the Prussian province of the Eiffel on the left bank of the Rhine are a considerable number of cones presenting very similar features to those of Auvergne, and open to the same remark.

In the island of Lancerote many such were thrown up by the terrific eruption of 1730, described by Von Buch, upon a straight line crossing the whole island. This chain of cones is represented as exactly analagous to that of Auvergne. At Jorullo, in Mexico, six cones were produced upon a line of this character by an eruption in 1759. (See Appendix.)

* See Plate I. (opposite the Title-page.)

in much greater quantities to the westward of the vents than on the opposite side, evidently from the influence of the *trade*, or constant east winds that predominate in those seas; and this observation, which, if correct in this case, may be extended to the disposition of the fragmentary products of all the inter-tropical volcanos of every age, will perhaps be found to have exerted no slight influence on the geology of these regions.

4. The line of projection may by some solid impediment be forced to swerve from the vertical, and a similar result would be produced to that of the last-mentioned cause.*

It is owing to the almost universal occurrence of some of the three last disturbing causes, that in volcanic cones of great

* The Puy de Chopine, near Clermont, remarkably exemplifies this circumstance—An enormous mass consisting of pre-existing rocks (mostly granite) appears here to have been heaved upwards by the first efforts of the eruptive force, and lodged in such a manner on its edge as completely to overhang the spot where the vent seems to have existed. Consequently, the ejected fragments have fallen and accumulated only on the opposite side of the orifice of eruption, so as to form a semicircular ridge called the Puy de la Goutte.

Fig. 10.

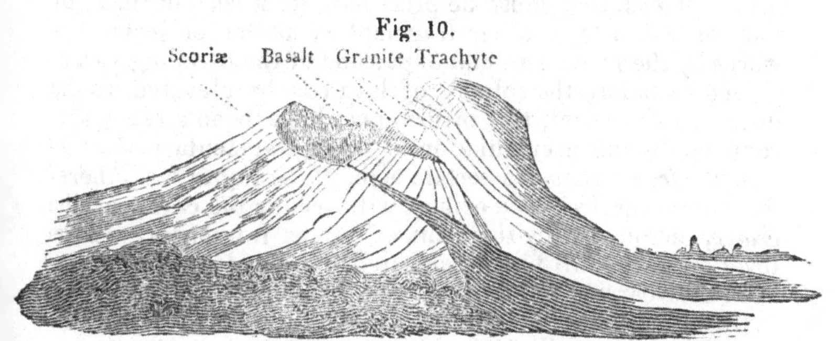

This ridge half encloses the extraordinary rock by which the dispersion of scoriæ in other directions was impeded, and which goes by the name of the Puy Chopine. The visible alteration of the granite close to the vent was obviously produced by its long exposure to the heated and acid vapours emanating from thence. The great vertical bed of trachyte which backs the rock on the opposite side, apparently rested upon the granite previous to the eruption; and the probability of this idea is strengthened by the fact, that rocks of trachyte are visible in situ on many points of the plain around the Puy Chopine, and attest the existence of a widely-spreading bed beneath the scoriæ that cover its surface.

regularity in other respects, the ridge seldom or never preserves a uniform height thoughout its whole circuit, but generally rises on one or more points above the remaining part. Hence those bosses which frequently show themselves on one or on opposite sides of the cone, and which from the gradual depression of the intermediate parts of the ridge communicate to the profile of the hill that saddle-shaped outline which has been often remarked as characteristic of volcanic mountains.

5. The last and principal cause which tends to disturb the regular accumulation of fragmentary substances around a volcanic vent, consists in the emission of liquid lava *en masse* from the same orifice. It has been already observed that the intumescence produced in the lava immediately below the vent, by the sudden reduction of the pressure it previously endured, must elevate it to the lips of the fissure, if the predominance of the expansive continues a sufficient time; and may often impel it, in greater or less quantity, on the outer surface of the earth, according to its fluidity, and the continuation of the intumescence.

But in all cases, the discharge of lava en masse from the orifice of eruption must be preceded, for a certain time, by that of scoriæ and other fragmentary matter projected upwards by the rapid rise and expansion of the escaping steam. Hence, before the column of lava can be elevated to the opening of the vent, this orifice must have been already surrounded by the accumulation of these fragmentary matters into a *volcanic cone* of a certain size. The liquid lava, therefore, on rising from the orifice, will occupy the cavity of the *crater*, surrounded by this bank of ejected fragments, and its level must be still further elevated more or less within this encircling barrier before it can escape laterally.

But since this barrier consists solely of incoherent materials, one of its edges will frequently yield to the lateral pressure of the body of lava occupying the crater, which instantly rushes out in torrents through the breach. Sometimes the whole side of the cone is, in this manner, broken down to the level of its foundation by the escaping lava, and the remaining segment loses proportionately in the regularity of its figure. In other instances the breach has not proceeded so far in depth; and only a part of the upper circumference has been carried away, producing a sort of notch in the ridge of the cone. Not unfrequently the walls of the crater appear to have been so solid, owing to the acuteness of the angle of their rise, or other causes, as to resist completely the pressure of the interior mass of lava: which could then only escape by rising to the level of the lowest part of the ridge, and pouring thence over the lip of the crater; like any liquid overflowing the sides of its containing vessel.* The peculiar conduct of the lava in the first of these cases, it is obvious, chiefly influences the *figure* of the volcanic cone, breaking down and carrying away one of its sides. In the other cases, it modifies the composition and structure, as well as form, of the hillock; for the lava, escaping from the crater either by a shallow breach, or over the lowest lip of its circumference, flows down the outer slope of the hill; and a considerable part of it, congealing and consolidating in this descent, remains fixed there as a solid and rocky prop or rib to the fragmentary cone; by which its general solidity and resistance to subsequent causes of destruction is materially augmented.

The residuum of lava left within the cavity of the cone, and which congeals superficially, so soon as the diminished energy of the expansive process checks the further protrusiod of any fresh lava over the lip of the crater, has the same effect.

But it has been seen above that the expulsion of lava from a volcanic orifice attests the crisis of the eruption, occurring during its period of greatest energy, and that this emission tends rapidly to weaken the forces of eruption, and bring on the decline and finally the termination of their outward activity. It has been also observed that after the lava has ceased to be propelled, the ejections of scoriæ, and other fragmentary matters, must usually still continue for a certain time, with a progressive diminution of violence, previous to the final cessation of the eruption.

It might, therefore, be expected whenever the expulsion of lava in currents had broken down or otherwise degraded a volcanic cone, that these subsequent ejections would repair it, and somewhat restore the regularity of its figure. This

* Instances of these different modes of procedure are numerous, particularly among the cones of Auvergne and the Velay. The Puys de la Vache, las Solas, Vichatel, Mongy, and Chaumont, are remarkable examples of the first case.—Le Puy Noir, and Louchadière, near Clermont, of Tartaret near Murol, of la coupe d'Ayzae in the Vivarais, &c., of the second.—Graveneire and Nugère amongst the Chaine des Puys, and many of the volcanic cones of the Prussian frontier of the Eiffel, of the third: (See Plate I. Frontispiece, fig. 1.)

may be thought indeed to have been very frequently the case, from the numerous instances remaining of tolerably regular cones, from the foot of which lava currents appear to have issued. In these cases it is evident that the cones were thrown up *after* the emission of the lava, and consequently conceal its source.*

Occasionally it is easy to distinguish the products of these subsequent ejections from those of the earlier explosions that preceded the emission of lava.†

Frequently, however, the cone remains in its breached and

---

* I may quote as examples of this common occurrence the Puys du Petit Dome, De Come, Pourcharet, Montchal, Montjughat, Lamoreno, and Chaumont, near Clermont.—A great proportion of the volcanic hills of the Velai are of this character, as well as many of the parasitical cones occurring on the flanks of Ætna, the Peak of Teneriffe, Palma, Lancerote, and probably of most large volcanic mountains.

† Thus for example the Puy de Pariou (Auvergne) presents the remains of the cone thown up by the first explosions of this eruption in a semicircular ridge, breached on one side by a current of lava, which is seen to have issued from it, and flowed to a great distance over the neighbouring plain. Upon this imperfect cone a second of excessive regularity was raised by the subsequent ejections. This last still remains wholly undisturbed by any posterior emission of lava, and contains an elliptical crater of considerable depth and great beauty.

See the plan and section of the Puy Pariou, figs. 11 and 12.

Fig. 11.

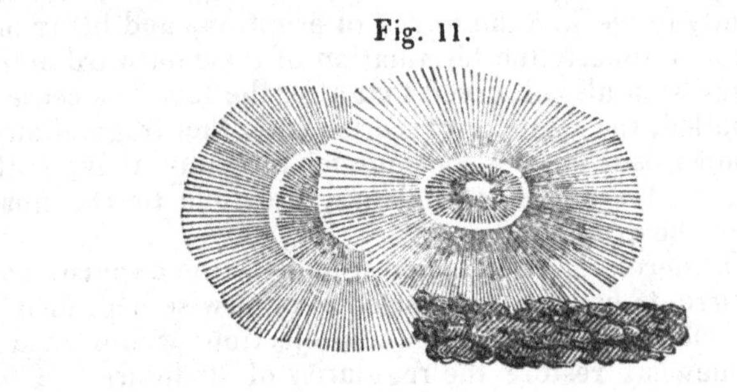

Fig. 12.

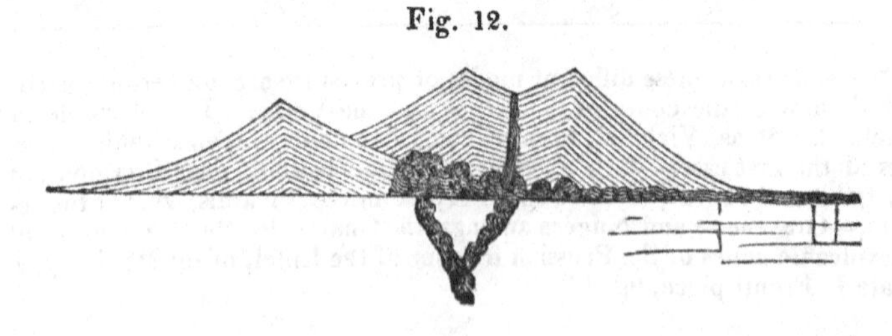

imperfect state, proving the fragmentary ejections subsequent to the expulsion of the lava to have been inconsiderable.*

This apparent anomaly may probably be occasioned by the weight and rapid congelation of the lava produced in the interior of the crater wholly repressing the disengagement of the elastic fluids, and putting a stop to the eruption soon after its energy began to diminish. But in many instances there is reason to suspect that the latest gaseous explosions found a vent from some other proximate point of the original fissure; that by which the lava issued being choked up, by its abundance, the consolidation of its surface, and the weight of fragments previously ejected. In this case a subsidiary cone will have been produced in the vicinity of the former one, and will be distinguishable from the latter by its undisturbed form, and the absence of a lava current. These latter cases are obviously most likely to occur where the lava has filled the crater without breaking down any part of its circumference, since the repressive force, occasioned by the weight of the mass obstructing the vent, is then at its maximum. If, notwithstanding this, any explosions of vapour take place after the lava has ceased flowing, the residuum which occupied the crater will be blown into the air, its fragments scattered on the outside of the cone, and a cavity worked out through the axis of the cone, the sides of which will be probably steeper than those of an ordinary crater; and in lieu of the shelving talus of loose scoriæ will present a circular wall of dislocated lava rocks.†

§. 4. Having thus far examined the varieties of form incident to a *simple volcanic cone*, or, "Hill produced by the ejection of fragmentary matter by a single eruption,"—it remains to observe upon its character of composition and *internal* structure. It will evidently have the aspect of a conglomerate consisting of the loose substances thrown out.

These are of various kinds, and may be classed in the following manner, viz.

---

* E. g. Puys de las Solas Vache, Noir, Louchardiere, &c. (See Plate I. fig. 1.)

† This appearance is frequent among the volcanic cones which stud the elevated mountain range that separate the Valley of the Allier from that of the Loire, towards the source of both rivers, in the Department de la Haute Loire.

It is observable also in some of the cones of the Eiffel.

1. Scoriæ, or portions of lava torn from the surface of the internal column by the exploding vapour, in a still liquid state, and consolidated rapidly by the sudden diminution of pressure they sustain on being launched into so rare a medium as the air; owing to which they are instantly and violently inflated by the expansion of the elastic fluids they contain, and rendered more or less *vesicular*. The greater part of their caloric is absorbed by the vapours thus expanded, and the remainder is rapidly parted with by radiation to the cooler atmosphere, through which these masses are driven; and thus, though they are seen to be projected in an incandescent and semifluid state, they quickly lose their brightness as they ascend, and fall to the earth in the shape of spongiform and tattered fragments, which are often vulgarly called volcanic *cinders*, from their resemblance to the scoriæ of our furnaces.

When the lava is possessed of any considerable degree of liquidity, in which case it always becomes proportionately ductile and glutinous, these fragments frequently assume a diversity of irregular forms, more or less fantastic, and occasionally pseudomorphous. Among the latter may be mentioned those of twisted cables, branches of trees, stalactites, and in particular pear-shaped or almond-shaped nodules. These last owe their figure to the violence of their projection in a liquid state; the resistance of the air, and their sudden and superficial condensation, impeding the expansion and escape of their confined vapour, and moulding them into a somewhat globular form, by the same effect which produces patent shot in our manufactories.

These rounded drops, or bombs, (*Bombes, larmes volcaniques*) often contain in their nucleus a fragment of some preexistent rock enveloped by a coating of lava; the fragment, having apparently fallen into the bath of liquid matter from the sides of the fissure of eruption, was projected into the air by the next explosion. The ordinary scoriæ thrown up by a volcanic vent usually bear an exact resemblance to those which form on the surface of the current of lava produced by the same eruption, at any distance from its source. But the most striking of these imitative forms, particularly the pear or bomb-shaped, can only be assumed by masses of lava projected liquid to a considerable height in the air; consequently the occurrence of a number of these upon any point proves it to have been the site of a volcanic vent; and this indication, of unquestionable evidence, may afford a very serviceable assistance towards recognising the situation of extinct volcanic vents, where their products have suffered much degradation from time or outward injury.

These nodular bombs being generally more solid than the lighter cellular scoriæ, resist the wasting influence of the weather and other agents of destruction, much longer and more effectually; and hence, it is not unfrequently the case, that, in districts of ancient volcanization, the site of an active vent, and consequently the source of a current of lava, may be recognized with absolute certainty when scarcely any thing remains of its original cone.*

The structure and external shape of the ejected scoriæ is influenced chiefly by the liquidity of the lava and its specific gravity, as well as by its temperature, the force of projection, and the density and temperature of the external medium into which it is thrown up.

An excessive liquidity joined to a low specific gravity, produces a viscous or glutinous consistence, and the scoriæ are drawn out into long vitreous filaments; or are inflated by the production of vesicles separated by thin glassy walls: the bubbles of vapour remaining confined within the mass, whose outer surface congeals with great rapidity into a thin but close crust that effectually impedes their escape. (Pomice, scoriæ of glassy lavas, &c.) The same degree of liquidity joined to a high specific gravity, gives rise to a largely cellular or cavernous structure; thick, solid, and fine granular walls separating the cavities; the greater part of the expanded vapour having effected its escape. (Basaltic scoriæ.)

An inferior degree of liquidity gives a massive, rough and fractured aspect to the ejected scoriæ, which contain only small, irregular, and generally angular cells; while a still lower liquidity altogether prevents the projection of scoriæ: the intumescent lava issuing alone from the orifice of eruption, and still inclosing all the vapour that has been developed from the water which it contained in a solid state.

Since the circumstances of temperature and external pres-

---

* The diminutive tumulus at the source of the current of basalt at Pradelles, on the right hand of the road from Clermont to Pontgibaut, near la Barraque, is an instance in point; and there are many similar ones in Auvergne, where the numerous volcanic products of different ages exhibit repeated examples of every possible gradation of the destroying process.

sure that accompany the consolidation of scoriæ or any other superficial mass of lava congealed by sudden exposure to the air, can seldom vary to any very influential extent, the liquidity of the lava in this condition will depend entirely on the comminution of its crystalline particles; and it may be therefore assumed as a general axiom, that the size and regularity of the air-bubbles will vary directly with the fineness of grain; while, on the other hand, the escape of the bubbles formed depending upon the specific gravity of the lava, the quantity of such cells remaining will vary inversely with this character.

From these observations it is seen at once, that the external structure of scoriæ must differ according to their mineral composition, since their specific gravity varies with it. And in reality we observe the scoriæ of felspathic, or *light* lavas, are highly vesicular or filamentous. (Pomice.)

Those of the heavy, or basaltic lavas, are proportionately solid, compact, and dense, with fewer or at least smaller pores, in direct proportion to their weight.

The fine glassy threads produced by the explosive escape of vapour from the felspathic lavas, when disintegrated to a degree of comminution amounting, or nearly so, to a complete fusion, sometimes, though rarely, descend in this form, resembling delicate spun glass. (Isle of Bourbon.)

In general these viscous filaments are twisted and rolled up together, in their ascent, into balls, or irregular masses which imitate a tangled skein of coarse silk. (Silky filamentous pomice of Lipari, Mt. Dor, &c.)*

On the other hand, in the heaviest or basaltic lavas, from which the vapour rises and escapes in explosions, by reason of its very inferior specific gravity, in opposition even to an

---

* The glassy coating of the air-bubbles of the coarser and heavier lavas appears to be owing in part to the rapid disintegration of the particles that form the sides of the vesicles by the friction of the moving bubble and its expansion; and perhaps, also, in great part, to such an arrangement of the minute crystalline particles that they present their flat surfaces to the interior. This would necessarily be effected by the expansive force of the confined vapour pressing upon them in the direction of the radii of the globule. Where the grain is coarse, the elasticity of the vapour had not power to alter the arrangement of the surrounding crystals, which consequently are found to project their sharp, and often very perfect, angles into the cavity. Where it was fine, it may be perceived by means of a lens, that the minute crystals have usually their facets arranged in a direction tangential to the radii of the globule. It is possible also that a certain degree of superficial fusion may have been produced in the lining of these cells by the heat transmitted to it from the proximate parts of the lava.

extremely imperfect liquidity occasioned by great coarseness of grain, and perhaps a lower temperature, the ejected scoriæ consist of dense crystalline fragments, ragged and scoriform, but containing few or no cells, which being triturated by their mutual percussion as they ascend, give rise to showers of isolated and often very perfect crystals of augite or leucite. (Stromboli, Monti Rossi on Ætna, Monte Albano, &c.)

The scoriæ, of all kinds, are also more or less broken, comminuted, and even pulverized, by the mutual friction they exert on one another in the air, and particularly by the repeated projections to which many fragments are subjected; falling directly back, or rolling down the slope of the crater into the vent, to be again thrown out by the next explosion; consequently they are found in fragments of all sizes.

When the average volume is small, the conglomerate is called *Lapillo*, or *Pozzolana*, according as it consists of felspathic, or augitic lava. When the trituration has proceeded still further, its result goes by the name of volcanic *sand*, or ultimately *ashes*.

II. The second class of fragmentary substances ejected from a volcanic vent by an eruption, and which enter into the composition of the cone produced, consist of fragments torn from the sides of the fissure of eruption by the expansive force of the aeriform explosions.

The rocks through which the fissure was broken, and consequently their fragments, may be of any nature, primitive, secondary, or tertiary; and even of a great variety of characters, belonging to different strata across which the vent was forced.

Thus the volcanic cones of the Auvergne and Velais, which are based upon what are commonly called *primitive* rocks, invariably contain numerous fragments of granite, gneiss, and mica schist; as well as of limestone when the vent opened within the limits of the freshwater limestone basin of the Limagne.

In the Eiffel, the fragments derive from the supporting formation of Grey-wackè slate.

Those ejected by Vesuvius are sometimes of granite or gneiss, much more frequently of secondary limestone, and occasionally of tertiary sandstone, or marls, replete with shells—the occurrence of these latter fragments is the more remarkable, because this formation, with the exception of one small spot in the isle of Ischia, does not show itself outwardly any where nearer than the vicinity of Rome.

The continuance of the same formation beneath the volcanic products of the Western coast of Italy is thus proved by these accidental fragments.

Such fragments of pre-existent rocks almost always bear signs of having suffered a certain degree of alteration, during or previous to their eructation from the volcanic vent. Many, as has been already mentioned, are found encrusted with lava—others are partially fused on the exterior or throughout; the elements of some appear even to have entered into new combinations after their decomposition by the volcanic heat.

Some volcanic conglomerates contain numerous fragments of rocks differing entirely from any yet discovered in situ on the surface of the earth, and containing many minerals never observed elsewhere. (Somma, Monte Albano, Monti Cimini,) lake of Laach, &c.*

The proportion of this second class of fragments to the former (scoriæ) in a volcanic conglomerate, must depend considerably on the solidity of the pre-existent strata through which the eruption broke forth. When these are composed

---

* It is never doubted that these fragments were torn from pre-existing strata through which the volcanic vent was forced; but it is not generally agreed upon by geologists, whether, 1stly, these strata are of the ordinary kinds, secondary and primitive limestones, gneiss, granite, &c. and that their fragments have been altered by exposure, perhaps for a long period, to the volcanic heat communicated by the neighbouring lava, and perhaps to other modes of influence, by which their elements may have been decomposed, to be re-aggregated again in different proportions, &c.; or, 2dly, that the strata broken through by the eruptions are uniformly of this anomalous character, and different from those that show themselves any where in situ on the surface of the earth. The latter opinion presents a striking improbability, while the former is unopposed by any argument, and is rendered highly credible by other analogous observations—such is the change in situ of secondary limestones to crystalline dolomite by contact with volcanic lava rocks, already observed in the Tyrol, Scotland, Ireland, Ponza, and other points, and particularly by the appearances presented in a great number of fragments of ancient lava currents which were ejected by the great eruption in October, 1822. These were evidently of Augitic basalt, similar to those which compose the greater part of the mountain; but this rock seemed, in these fragments, to have suffered a high degree of torrefaction, probably by long continued exposure in the vicinity of the main vent to the volumes of intensely heated lava by which it had been long filled. Parts of these rocks were fused into a basaltic obsidian; on some points the leucites alone were fused; in others they were rendered carious or had entirely disappeared, their cavities being occupied or coated by numerous beautiful crystals of melanite, and by delicate acicular crystals of an unknown mineral which had evidently been formed from some of the elements of the leucites subsequent to the consolidation of the rock.

of granite, limestone, or other compact rocks, the quantity of their fragments broken from the sides of the fissure, and ejected, will be naturally less than in a district of schistose and arenaceous rocks, which are of a looser texture and more fragile. Many of the volcanic cones of the *Eiffel*, whose eruptions chiefly broke through a formation of grey-wackè slate, are almost wholly composed of fragments of this rock, more or less calcined, *never fused.* Their proportion to those of scoriform lava is frequently as much as nine to one, while among the cones of the Auvergne based upon granite, the fragments of this rock are very thinly scattered. Again, pieces of mica-schist and gneiss occur more plentifully in the Velay, these rocks being more fragile than granite.

It is obvious that fragments of this class will very rarely occur amongst the ejections of the later eruptions of an ancient habitual vent.

Thus, while they are found abundantly in the conglomerates of Somma, they seldom or never are thrown out by Vesuvius; and even when this happens, they may be suspected to have previously formed part of some of the conglomerate strata of the old cone, broken up by the lateral extension of the avenue of eruption.

The finest and lightest of these fragmentary ejections are borne upwards to a great height by the torrent of heated and aqueous vapours that escape from the volcanic chimney, and mingled with these in thick clouds, remain suspended in the air for a considerable time, and are often carried to great distances by atmospheric currents before they fall to the earth in showers of ashes. The immense friction they sustain is probably the cause of the great development of electricity that often accompanies their projection.*

The heavier fragments describe paths of various curvature in their ascent and fall according to their volume and weight, and the force and direction of the impulse communicated to them by the violence with which the elastic fluids escape from their subterranean confinement.

---

* The electrical phenomena that accompanied the eruption of Vesuvius in October, 1822, were singularly beautiful. From every part of the immense cloud of ashes which hung suspended over the mountain, flashes of forked lightning darted continually. They proceeded in greatest number from the edges of the cloud. They did not consist, as in the case of a thunderstorm, of a single zig-zag streak of light; but a great many coruscations of this kind appeared suddenly to dart in every direction from a central point, forming a group of brilliant rays resembling the thunder-bolts placed by the ancient artists in the hands of the cloud-compelling Jove.

It is evident that the impulsive force acts most powerfully in the vertical direction and from the centre of the vent—the more the line of projection diverges from this, the less must be its comparative energy. Again, the divergence of the ejected matters must diminish proportionately to the depth within the volcanic vent of the surface of lava from whence the explosions burst, and on the contrary will be at its maximum when this surface is highest—that is, even with the lips of the orifice.

The height to which fragments, even of considerable volume and weight, are carried vertically by the maximum of the projectile force is often surprising, and conveys a striking impression of the powerful tension and consequent elasticity of the escaping fluid.*

To account for this immense projectile force it need only be remembered, that the confinement occasioned by the walls of the vent, and the lateral pressure of the lava, must concentrate all the explosive force of the expanding fluids into the direction of their escape, acting upon them exactly as the barrel of a gun on the gasses generated by the combustion of powder in its breech. The same consideration explains the powerful effect of the expanding vapour in breaking up and enlarging the fissure of eruption when the surface of the lava is at some depth within it, just as the barrel of a gun bursts by the ignition of a large charge, if the length be out of all proportion to the bore.

§. 5. The substance of the simple volcanic cone will therefore be a conglomerate, consisting of the fragmentary matters of one or both of the two classes mentioned above. This conglomerate will be either wholly amorphous, or composed of more or less distinct beds, according as the falling substances varied at intervals in their general characters of size, colour, mineral nature, form, &c.; in the same manner as a mass of alluvial soil is distinguishable into beds by variations of character in the drift matter successively deposited.

These beds, or strata, will present a characteristic arrangement resulting necessarily from their mode of deposition.

---

* Vesuvius has been seen to vomit scoriæ to a height of nearly 4000 feet. Cotopaxi has projected a mass or rock of 1000 cubic feet to a distance of three leagues.—*La Condamine*, *Voyages l'Equateur*.

The same volcano sends up occasionally showers of fragments to a height of more than 6000 feet above its crater, which is itself nearly 19,000 feet above the level of the sea.

In fact, if we consider the figure described by the *ridge* in its progressive elevation during the formation of the hillock (and which abstracting all causes of irregularity will approximate to the superficies of an inverted conical truncation), we shall find the cone is divided by it into two segments, the strata of which will, in the case of the inner segment, dip inwardly, or towards the axis of the cone;—in that of the exterior segment outwardly, at nearly the same angle to the horizon (which angle will usually oscillate about 25°), according to the size and coherence of the component fragments.*

The planes of the strata in either case will also be parallel, or nearly so, to the external surface, or slope of the hill, unless this has been subsequently degraded, and will have a circular or elliptical direction, with a degree of curvature dependent on the magnitude of the cone and their respective distances from its axis—that is from the vent of eruption.

Fig. 13.

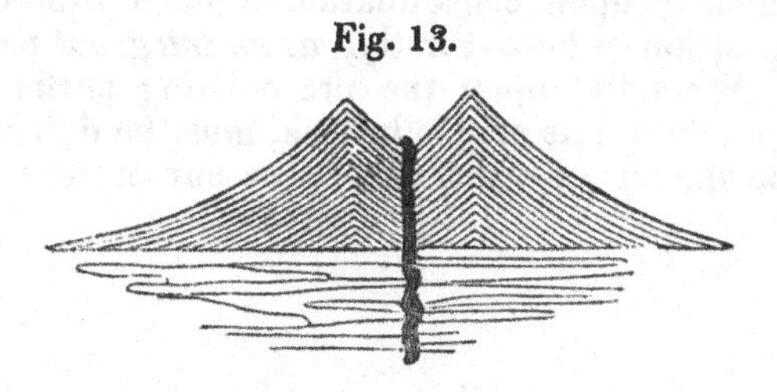

Such a peculiar mode of arrangement eminently distinguishes those stratified conglomerates that were deposited in this manner as part of a volcanic cone, from fragmentary strata of any other origin, volcanic or not; and this consideration becomes of great service in the examination of extinct volcanic regions, where the cones have been much degraded by subsequent mechanical injury. Even where all regularity of external figure is wholly lost, and perhaps only a fragment of the former hill remains, still the internal structure will often be sufficient not only to disclose its original character, but to point out the exact position of the vent from which it

* The promontory called *Capo di Miseno*, near Naples, presents a natural section, effected by the erosion of the sea-waves, of a simple volcanic cone, one of those thrown up from the numerous vents of the Phlegræan fields. The structure and stratification of this cone is precisely such as would be expected from the reasons in the text.

was thrown up, and consequently the source of the neighbouring currents of lava, and the direction of the subterranean fissure; facts of considerable geological importance in the study of such districts.

It must however be remembered, that the disturbing causes enumerated above, as modifying the outward figure of a volcanic cone, have an equal influence on its internal structure. All the varieties of dip, direction, and comparative thickness, that cannot fail to be produced in its conglomerate beds by these several circumstances, will be made obvious by the simple consideration that the disposition of the different strata coincides with that of the superficies of the cone during the successive instants of its gradual increase—the one must, therefore, have always varied with the other.

When any quantity of *lava* was protruded from the vent in a liquid form, this circumstance will have influenced, not merely the form and structure, but the *composition* also of the cone, producing upon consolidation a hard lithoidal rock, whose disposition either—Firstly, *as an integrant* part of the cone; or—Secondly, upon the pre-existing surface of the soil, without the limits of this hillock, must be determined by causes, the nature of which it will be our present object to examine.

CHAP. IV.

Conduct of Lava when protruded on the Surface of the Earth.

---

§. I. Before we enter on this investigation it may be as well to premise a few general remarks on the mineral characters of the different rocks which are recognised to have been produced as lavas from the interior of the earth, for the purpose of facilitating the illustration of the laws relating to their conduct which may be deduced from our theory.

The principal constituent minerals of these rocks are felspar, augite, and titaniferous iron; and they are usually classed by mineralogists into two great families, according to the prevalence of the first, or the two latter of these minerals; or of their equivalents.

Those volcanic rocks which are composed almost entirely of felspar, with occasionally some crystals of quartz, and more frequently of mica (augite and titaniferous iron being either wholly absent, or not appearing in any greater proportion than five or six per cent.), are denominated *trachyte.*

Common trachyte has generally a coarse granitoidal grain; and a harshness of texture, owing to its porosity and the irregular disposition of its component crystals. When the crystals, or at least the largest of them, lie in parallel planes, so as to give a laminar or slaty structure, the rock is called Phonolite, or Clinkstone.

When the grain is very fine and compact, but at the same time irregular and unequal, it is called Trachytic Porphyry: when this fineness of grain is joined to a scaly structure—Compact Felspar. When the crystalline particles have been comminuted nearly or quite to the extremest degree, so as to give either a resinous or glassy texture to the rock, it is denominated Pitchstone and Obsidian.

On the contrary, where either augite or titaniferous iron, or both, occur in so large a proportion as to give their general characters of colour and weight to the rock, it is termed *basalt.* In this rock olivine usually makes its appearance, and the felspar is occasionally replaced by leucite, and the common augite by hornblende (hemi-prismatic augite—*Jameson*), in very rare instances by hauyne.

Those basalts in which the grain is extremely coarse, appearing to consist solely of visible crystals of augite, are termed Dolerite. When hornblende is substituted in this coarse mixture for augite, the rock is called *Greenstone.*

The finer grained compounds of either sort are named Common Basalt. When the structure is slaty, from the parallelism of the crystals either of felspar or augite, it takes the name of Basaltic Clinkstone. When reduced by extreme disintegration to a glass, it has been called Gallinace or Basaltic Obsidian.*

Now the specific gravity of felspar is to that of augite, hornblende, and titaniferous iron, the other principal constituents of these rocks, in the average proportion of four to five. Consequently the specific gravity of a lava will vary directly with the proportion in which it contains the heavier minerals augite, hornblende, or titaniferous iron, which probably owe their superior weight to the quantity of iron that enters into their composition—and the two principal classes of lavas, the felspathic and ferruginous, producing severally on consolidation trachyte or basalt, might with propriety be distinguished as the *light* and the *heavier* lavas.

The *fluidity* of a lava, or the facility with which it moves in any or all directions in obedience to the impulse of its own gravitating force, is compounded of its liquidity, or the mobility of its parts, and its specific gravity.

Consequently, with an equal degree of liquidity, the fluidity of any lava will vary with the proportion of the heavier or ferruginous minerals in its composition.

---

* I am of opinion that it would be found useful to institute a third and intermediate class of rocks, which should include those compounds in which the proportions of felspar and the ferruginous minerals are so nearly balanced that the general aspect of the rock is neither entirely felspathic nor ferruginous, and that it appears to have an equal right to belong to either class. These rocks might be called *greystone*; their prevailing and I believe constant colour being some shade of grey. This class would nearly correspond to the tephrine of de la Metherie and Cuvier, and would include many clinkstones. The lava of Volvic, and that used as building stone at Mont Dor les Bains, may be given as examples of it.

But the liquidity of a lava under the same circumstances of pressure and temperature varies as we have seen (p. 27) with the average comminution of its crystalline molecules. Hence lavas of the same mineral quality and therefore of equal specific gravities, when produced under similar circumstances of temperature and pressure, will possess a degree of fluidity inversely proportionate to the size of their grain. And, vice versa, of lavas in which the grain averages the same degree of fineness, the fluidity should appear to have been directly proportionate to their specific gravity.

These axioms must be attended to, because, if by observation we should find evidence of their correctness it will be tantamount to a proof of the truth of the hypothesis from which they are deduced.

§. 2. It has been seen that in all cases of volcanic eruption the intumescent lava rises to the mouth of the vent, or to such a height within it as allows of its free communication with the atmosphere; and that it either remains at that level, the aqueous vapour escaping in bursts from its surface, as fast as it is generated below in the focus of expansion; or overflows the edge of the vent, and is discharged on the surface of the earth, in greater or less quantity according to its liquidity and specific gravity, (i. e. its fluidity,) and the energy of the eruption, i. e. the quantity of ebullition that takes place in given times in the subterranean focus.

As it is elevated the lava becomes proportionately dilated, entering still farther into ebullition as the pressure upon it diminishes with the length of the column of liquid it has to support, and which acts both by its weight and the reaction of its elastic vehicle from the lateral resistances which oppose its entire expansion.

When any portion of lava has reached the mouth of the vent, and comes into contact with the outer fluid medium, both these repressive forces are reduced to nullity, and the sole remaining resistance to the full expansion of the steam it contains consists of the pressure and density of the overlying medium. Consequently the steam is rapidly dilated, and escapes in great quantity from the exposed surface. But by this expansion the temperature of the surface is instantly lowered, the vapour that remains proportionately condensed, and the solid crystalline particles of the lava brought so closely together that their forces of attraction prevail over the weakened force of expansion, and they reunite into a solid mass. The superficial crust that instantly hardens in this manner on the exposed surface of any lava will impede the expansion and escape of more vapour from beneath it, in a degree proportioned to the closeness of its texture. When this is at the maximum, the first film that hardens on the surface will offer an impenetrable barrier to the escape of the vapour from below, and prevent, for some time at least, the progress of the consolidation downwards. On the contrary when the lava is porous the steam of the interior will be enabled to percolate through the outer crust in obedience to its expansive force, and the greater the porosity of the lava, that is the more numerous and large interstices, the more easily and rapidly will this escape be effected. Consequently the depth to which the process of consolidation in this manner extends inwardly from the exposed surface of a lava in given times will be inversely proportionate to the compactness of its texture,* which must itself vary inversely with the size and irregularity of its component crystalline particles.

This law by which the porosity and compactness of a lava is determined will be further investigated in a subsequent chapter.

In proportion as the crust that hardens in this manner on the surface of lava, checks the expansion of the vapour beneath, it continues the pressure sustained by the liquid, and even augments it by the contraction which accompanies the process of congelation.

But the reaction of the expansive force of the compressed steam, together with the effect of this contraction, generally splits and breaks up the hardened crust. The fissures formed by this shrinking allow of the expansion and escape of more steam from the interior, and this gives rise to fresh cracks; and in this manner the consolidation is propagated to a considerable depth, while the surface becomes more or less broken, rugged, and uneven.

In this manner, when a body of lava is exposed to either of the external fluid media, its surface becomes instantly congealed;—the depth to which the induration proceeds varying directly with the size of its grain and the irregularity of its arrangement.

Below the point to which the induration has reached in obedience to this law, the matter of the lava remains long in a state of liquidity, since it can only lose this character by parting with the caloric it contains by gradual transmission

* From a similar cause it is that coarse clay desiccates inwardly sooner than fine grained, the water which forms the vehicle of its liquidity escaping with the greater facility in proportion to the porosity of the resulting solid.

through the slow conducting powers of the crust. In this condition it obeys of course the laws that regulate the motion of all liquids, and if urged to move in a lateral direction by any force, it instantly acquires a similar crust or solid coating upon every fresh surface it exposes in succession.

It is therefore only at the extremity or borders of a moving mass of lava, that it appears to be in motion, or is visibly liquid and incandescent, and even there only when the movement is tolerably rapid. The rough superficial crust, which forms instantaneously on its exposure at any point, remaining stationary, or nearly so, and allowing the lava to flow on from beneath it.

§. 3. The disposition assumed by any body of lava protruded on the surface of the earth by volcanic expansion must obviously be determined by the compound influence of the expulsive force, of its fluidity, and of the various external circumstances that may modify the permanence of its fluidity, or that tend to favour or impede its progress in a lateral direction.

I. The force of expulsion, by which fresh volumes of liquid lava continue to be driven forth from the vent in greater or less quantity, might rather be termed the ratio of production, since it varies directly with the quantities of lava produced in given times.

This ratio, when the *fluidity* of the lava is fixed, varies as we have seen directly with the rapidity of the internal progress of expansion through the focus, i. e. with the energy of the eruption.

If this term be fixed, it varies inversely with the fluidity of the lava, since the degree of intumescence by which it is occasioned preserves this proportion.

This law however, it must be remarked, is only applicable when the vent of emission corresponds to the summit of the column of lava elevated within the volcanic chimney.

Whenever, as will appear hereafter to be a frequent case, by the sudden formation of a vent or fissure of escape through the side of a volcanic mountain,—the lava is allowed to issue at a lower level, the force of its expulsion will be compounded of the weight of the column of liquid already raised above that level, and the energy of the expansive force. The influence of the force of expulsion on the disposition of the lava expelled, is evidently to increase its superficial spread by augmenting the velocity of its lateral progress.

II. We have already considered the circumstances by which the fluidity of lava is determined, viz.—its liquidity and specific gravity; and it has also appeared that the permanence of its fluidity, that is, the difficulty opposed to its congelation, varies directly with its fineness of grain, ceteris paribus. It is obvious that the direct effect of the excess of these qualities on the disposition of a body of lava is to augment the rapidity of its advance in a lateral direction, and consequently the extent of its spread.

III. The external circumstances that affect the disposition of lava upon the earth's surface are divisible into, First,—those which tend to diminish its fluidity, and Secondly,—those which favour or impede the lateral extension to which it is urged by that fluidity.

1. The first class of circumstances comprehends the nature and condition of the external fluid medium with which its surface is brought into contact. The pressure sustained from the weight or elastic force of this medium, and the hindrance which it opposes to the escape of the steam according to its density, tend proportionately to diminish the fluidity of the lava. The latter difficulty may perhaps be diminished by the velocity of any currents that may traverse this medium, and by presenting continually fresh surfaces of contact favor the escape of its caloric and the emanation of its gaseous vehicle.

2. The external circumstances that tend directly to favor or impede the lateral extension of the lava consist principally in the superficial form and nature of the ground upon which the lava is produced. If this slope towards the orifice on all sides, the gravitating force of the lava opposes that of expulsion, and impedes its lateral expansion. If the surface be level, the lava must spread outwardly on all sides of the vent with a velocity, and to an extent determined by its fluidity and the force of expulsion. If the ground slope away from the source in any direction, the lava flows in that direction with a rapidity, and to an extent determined by the force of expulsion, its fluidity, and the degree of inclination of the surface.

In the same manner every obstacle against which the lava impinges in its course tends more or less to modify its direction and impede the velocity of its progress. But the effect of such impediments on the progress of a body of lava must be far greater than what they would produce on any other fluid in motion possessed of less elasticity or a greater degree of liquidity.

In this case they will more or less not only impede the lateral progress of the lava in the direction in which it was moving, but also diminish, and to a certain distance destroy, its fluidity, by increasing the compression on the contained elastic fluids upon which its liquidity depends.

In fact, while the lava moves on in obedience to its own gravitating force, and with a velocity proportioned to its fluidity and the inclination of the ground, it acquires a certain momentum, and when an obstacle in any degree opposes its progress, the reaction of this momentum from the resisting surface must occasion a great increase of compression on the lava immediately in its vicinity. By this the elastic fluids it contains, and which communicate to it the imperfect liquidity it possesses, are condensed, and the mass more or less solidified.

The partial consolidation thus effected is propagated backwards to a certain distance from the surface in proportion to the size and angle of the resisting obstacle, ceteris paribus.

The accumulation of the lava upon itself, as its velocity is stopped by any impediments towards the lower part of the current, tends also to solidify this part, by lengthening the column of liquid it supports. By both these causes united the fluidity of lava is more or less destroyed to a certain distance from every obstacle that checks it progress, and its advance must be proportionately impeded.

Upon the compound influence of all these circumstances will depend the direction taken by any body of lava on issuing from the volcanic vent—the velocity of its progress—the extent of its superficial spread—and consequently the figure and relative dimensions of the mass it forms on consolidation.

When those circumstances prevail that are favorable to the lateral progress of the lava, it will extend itself with a proportionate rapidity* over a considerable surface, and accord-

* A current of Vesuvius in 1776, flowed a mile and a half in 14 minutes; and M. de Buch witnessed one in 1805, which reached the sea in three

ing to the disposition of the neighbouring levels—either overflow a wide, and perhaps gently sloping plain; or spreading more in length than in breadth, perhaps occupying the bottom of a valley, or the narrower channel of its river, stretch like a continuous stream to a considerable distance from its source until the cessation of the force of expulsion, and the progress of consolidation in the mass already expelled, gradually diminish its velocity, and finally arrest its further progress.†

In the former of these two cases the bed of lava preserves on solidification the form of a *sheet*, (nappe) or *plateau.*—In the latter that of a *stream* or *current* (coulèe).

When on the contrary, the fluidity of a lava is very imperfect, and external circumstances unfavorable to its lateral extension, the lava must accumulate upon itself into bulky masses in the immediate vicinity of its source; and, according to the direction it is led to take by the neighbouring superficial levels,—either produce on *one* side of the orifice of protrusion a huge current or bed, of great thickness in proportion to its superficial dimensions, and to which we will give the name of a *Hummock*—or accumulating on *all* sides of the vent, one bed slowly spreading over another, so as to form a series of concentric coats, the mass will assume somewhat of the figure of a *dome* or bell—in which the vent occupies the axis, and will remain closed after the lava ceases to be protruded, unless broken through by the subsequent explosions of vapour.

§. 4. In adducing examples of these different modifications of figure assumed by a mass of lava after its emission upon the earth's surface, in obedience to the laws that regulate the motion of all substances in a similar state of semi-liquidity, —it will be difficult not to perceive a remarkable fact, viz. that the lavas which are mineralogically classed as basalts,

---

hours from the summit of the mountain, a distance of 3200 yards in a straight line.

† The dimensions of some lava currents produced even by recent volcanic eruptions are astonishing. Sir W. Hamilton reckoned the stream which destroyed Catania, in 1669, to be 14 miles long, and in some parts six wide.

Recupers measured the length of another upon the northern side of Ætna, and found it 40 miles; Spallanzani mentions currents of 15, 20, and 30 miles, (Voyage en Sicile, vol. i. 219.) and Pennant describes one which issued from a volcano of Iceland, in 1783, and covered a surface of 94 miles by 50! (North Globe, vol. i.)

from the prevalence of the ferruginous minerals—augite, hornblende, or titaniferous iron in their composition,—are almost universally found to have spread into thin *sheets*, or long and shallow *currents*, to a considerable distance from the orifice of protrusion;—while those lava-rocks which consist almost wholly of felspar (trachytes) are as uniformly disposed in massive beds, hummocks, or domes. Take for examples the extensive and shallow basaltic plateaux of the Mt. Dor and Cantal—of the environs of Le Puy en Velay—of the north of Ireland—of the plains of Iceland—and of the gentle slopes of the isle of Bourbon—take the innumerable long and straggling streams of the Auvergne—of Ætna, Vesuvius, &c.—and compare the conformation of these and other basaltic beds, with the massive trachytic hummocks that are closely grouped around the volcanic centres of the Mt. Dor, Cantal and Mt. Mezen in France, with the bell-shaped masses of the Puy de Dome, de Sarcouy, and de Cliersou, with the numerous trachytic domes of Hungary, and with the still more stupendous beds of the American Andes. Such a comparison presents a striking evidence of the truth of the observation.

It may be even asserted generally, that the greater the proportion of the ferruginous minerals (augite, hornblende, and iron) in the constitution of the lava, the greater appears always to have been its superficial spread, when poured out upon the earth's surface, other circumstances remaining to all appearance the same.

This so remarkable and constant relation between the mineral nature of a bed of consolidated lava and the proportions of its different dimensions, has been already recognized by Geologists; but many have been unfortunately led by this remark to the adoption of a serious error as to the origin of the trachytic and phonolitic rocks, which are considered by them as in no instance to have flowed on the surface of the earth, but to have been always elevated en masse into the position they now occupy.

Some writers have even gone the length of supposing that they swelled up like a bladder by inflation from below, (De Buch, Humboldt,) and are consequently still *hollow within*—a gratuitous supposition entirely at variance with all that we know for certain concerning the nature and mode of operation of the volcanic energy.

It is on the contrary obvious that the remarkable bulkiness of the felspathic lavas is fully and simply accounted for by their imperfect fluidity, which has been already recognised to diminish, ceteris paribus, with their specific gravity, and by no means induces the necessity of supposing any other mode of volcanic action than that by which the basaltic lavas were also produced. But it is likewise wholly and strikingly untrue that trachytes never occur in sheets or currents (nappes ou coulèes.) On the contrary when reduced to a great degree of comminution and possessing a high liquidity, the felspathic lavas have universally assumed that mode of disposition, as in the instance of the numerous streams of obsidian or glassy trachyte in Lipari, Teneriffe, Bourbon, Iceland, &c.; and under favorable external circumstances—as when the inclination of the ground from the orifice of protrusion was considerable, lavas of this quality, even though extremely coarse-grained, and therefore very imperfectly liquid, are frequently found to have spread laterally to a considerable distance, with this only distinction, that the currents thus formed are far thicker and more bulky than would have been the case with basaltic lavas of the same texture, and under the same outward circumstances.*

---

* To prove this assertion it were sufficient to instance the well-known trachytic current of the Solfatara, the numerous ones of the Isles of Ischia, Lipari, Palmaria, and Volcano, of Iceland, Guadaloupe, and the Isle of Bourbon, where both compact and earthy trachyte *appear in currents* (massive) flowing out of craters that still exist in their integrity. But the Geologists who uphold the contrary opinion by a strange perversity of reasoning refuse to class these rocks as trachytes; not from any observed difference in their mineral characters (which perfectly agree with those of some of the oldest trachytes) but precisely because they appear as currents, and have been obviously produced in the manner of ordinary lavas from a volcanic mouth. It might be remarked to these gentlemen, that they are arguing in a vicious circle—that the term trachyte is purely and essentially mineralogical as much so at least as that of granite, mica-schist, porphyry, limestone, or basalt—and that they have no more right to refuse this denomination to the recent rock of the Solfatara than that of granite to the secondary granite of the Val de Lavis, or porphyry, to the rock of this nature, which occurs in the red sandstone formation.

But to clear up all doubts on the subject, it may be as well to mention that many of their own *standard trachytes* are indubitably disposed in regular sheets and currents, and have flowed to considerable distances over the surface of the pre-existing beds. For example, in the vertical sections of the Mont Dor which are exhibited by the cheeks of its deep central ravines and gorges, many varieties of earthy trachyte may be seen forming thick but very extensive beds or layers superposed to one another to the number of five, six, or seven, and separated by intervening strata of tufa

At the same time that it cannot be contested that trachytic lavas have frequently flowed as currents and sheets (coulèes) on the surface of the earth, under circumstances of liquidity, inclination, and propulsion, peculiarly favorable to the velocity of their course and consequently the spread of the mass, yet it is certain that these currents are usually of very considerable thickness in proportion to their superficial extent, and approach more or less to the mode of disposition which we have designated by the term hummock, and which is in fact only a current of extraordinary bulk. It is likewise equally true on the other hand, that when all outward circumstances were remarkably *unfavorable* to the lateral progression of the lava, and its liquidity imperfect in the extreme, the protruded mass may have remained almost vertically above the orifice or fissure of eruption, scarcely spreading at all in a lateral direction and presenting the figure of a dome or hummock, which subsequent degradation will frequently have worn into a fan-

---

(or pomiceous conglomerate), occasionally even alternating with parallel beds of basalt. (Cascade du Mont Dor, Val, de la Cour, &c.) All these have beds a quâquâ-versal dip from the axis of the mountain, and a gentle and gradually diminished inclination, having evidently flowed from the lips of the central vent, (or crater,) whose remains are still observable in the deep circus, where the rivulets of the Dor and Dogne unite their waters without losing their names.

The great volcanic mountain of the Cantal exhibits numerous similar instances; or, indeed, it would be rather more correct to say, that here the trachytic lavas have almost all flowed to a considerable distance from the centre of eruption, producing immensely extended plateaux of considerable thickness whose super-position to conglomerates and other similar trachytic beds (as well as to fresh water limestone strata) is frequently observable in the steep flanks of the numerous ravines that stretch like rays from the centre of this conical mountain. The group of the Monti Cimini, near Viterbo, affords instances of the same disposition—trachytic currents based on tufa may be observed to form the slopes of this mountain.

There can be little doubt that the trachytic groups (volcanic mountains?) of Hungary would present many similar facts, if studied with a previous knowledge of the usual disposition of volcanic products. That they did not appear in this light to M. Beudant is perhaps attributable to this geologist having examined them rather too much in the spirit of the Wernerian school than with an eye to the laws that regulate the structure of a volcanic mountain. It is the more allowable to presume this, because M. Beudant himself asserts in the same positive manner, as the result of his own observations, the total absence of trachytic currents in the Mont Dor; where it is no exaggeration to say, that they are as evident as those of basalt, which are seen to compose the frame-work of Ætna or Somma, in the sections afforded by their deepest ravines.

tastic resemblance to ruined castles ("des chateaux-forts en ruines," Humboldt, Essai Geognostique.) It is also obvious that the complete dome or bell will only be produced when the aperture of emission is circular, or nearly so, and surrounded by a level surface. If the ground slope away from it, the dome will pass into a hummock, and by its extreme prolongation into a current. While on the other hand should the orifice be a longitudinal fissure, and the lava be produced on more points than one along its extent, by accumulating upon itself on either side of, and above this fissure, it will give rise to a lengthened hummock, which will scarcely be distinguishable from one that has flowed from a single source.

In all probability many of the massive trachytic formations of the Andes (to some of which M. de Humboldt gives the astonishing and almost incredible vertical thickness of 18,000 metres) were produced in this manner from wide and lengthened fissures disgorging during a long period of eruption from numerous vents upon their whole extent an immense quantity of felspathic lava, under circumstances peculiarly unfavorable to its fluidity.

There exists even in France, though as yet almost unremarked, certainly misunderstood, a colossal range of clinkstone perhaps capable of vying with many of the trachytic formations of the American continent, and at least fitted to raise the imagination to the conception of such enormous lava-streams. It has its origin in the Mont Mezen, (near le Puy en Velay) and stretches into the valley of the Loire, 30 miles off, with an average width of about six miles, covering therefore a surface of about 156 square miles. It may be seen to rest upon basalt in some places, and on fresh-water marles, and clays in others; more frequently still on granite. This, and its gradually inclined slope, from the heights of the Mezen into the ancient bed of the Loire, where it terminates, the extremity leaning against the foot of the granitic range of La Chaise-dieu on the opposite bank, prove it to have been produced rather as a continuous lava current, than from a lengthened fissure or in any other mode. The great comparative fluidity of the clinkstone lavas, which will be noticed immediately, in one direction, viz. coincident with the planes of their crystalline laminæ, favours this idea; as well as the enormous vertical thickness of the bed, which in many places reaches from 500 to 600 feet.

The atmospheric agents of erosion have effected great changes in this bed since its production, cutting entirely through it on some of the softer parts, and notching it into a number of fantastically shaped eminences, generally conoidal.*

The local differences of hardness and structure, by affording more or less facilities to the process of degradation, particularly to the action of frost and rain, and the easy destruction of those parts of the rock that were based upon friable freshwater marles and clays, to which the fissile structure of the clinkstone suffered the rain-water to percolate from above, were no doubt the causes of this extremely partial weathering.

The Puys de Dome, de Sarcouy, and de Cliersou, are beautiful instances of the dome or bell-shaped disposition, in its extreme perfection. These hills, the two last of which have been quarried to a considerable depth inwardly, present indications of a concentric foliated structure on the largest scale, that must necessarily have been produced by the peculiar mode of their production, one bed of semi-liquid lava lapping over the other as it slowly boiled over the lips of the central orifice.

Fig. 14.

Dome of Trachyte.

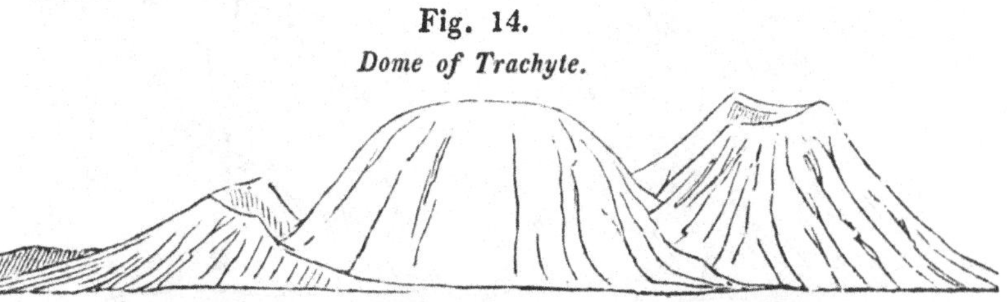

Profile of the Puys de Petit et Grand Sarcoury et des Goules.

Fig. 15.

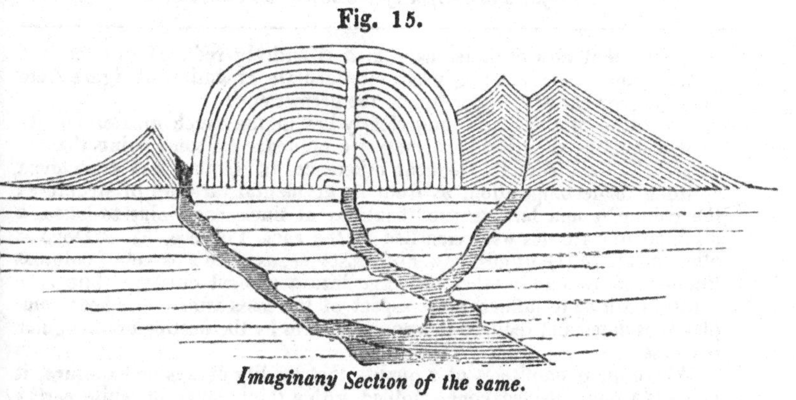

Imaginany Section of the same.

* MM. de Humboldt and Daubuisson have remarked the tendency of phonolitic mountains to waste into detached masses of a conical form. No where could the truth of this observation be better appreciated than along

The explosions of aeriform fluids which preceded or followed the emission of the lava, probably threw up the cones of fragmentary matters, which are so closely in contact with these trachytic domes (viz. le petit Dome, le Grand Cliersou, and le Puys des Goules.)

In a neighbouring instance, that of the Puy de Gromanaux, the subsequent explosions evidently broke through the centre of the trachytic dome itself, and left a hollow cavity or crater there, half-encircled by cliffs of trachyte, and containing a small but perfect cone, (Le Puy de Besace.) Another very complete example of the bell-shaped hillock of recent formation occurs in the Mamelon centrale, described by M. Bory de St. Vincent, as rising from the summit of the cone of the active volcano of Bourbon; this too has been hollowed out by subsequent explosions.

Fig. 16. Fig. 17.

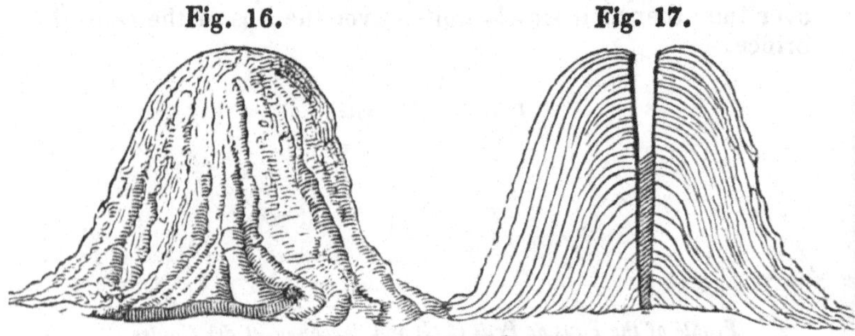

Profile and Section of the Mamelon centrale.

---

the range that now occupies us, which is entirely reduced to a series of rocky eminences, presenting every intermediate gradation of figure from the rude segment of a bulky bed to the perfect cone.

The cause of this uniformity clearly lies in the much greater facility with which this rock yields to meteoric influence on some points than on others, as well from its frequent differences of texture and consequent aptitude to decomposition, as from its accidental varieties of structure; the columnar and laminar modifications at times combining to hasten a disunion of parts (as was remarked in the rock Tuiliére, Mont Dor,) at others to afford the utmost power of resistance, as when a parcel of columns leaning against one another converge into a conical cluster. The same causes continue to influence the aspect of the mass after it has been completely isolated and reduced to a rounded form by the wasting of its angular portions.

Where the phonolite is of a quality that readily decays on exposure, it presents a smooth-sided cone, clothed with a thick layer of white earthy soil, which frequently supports luxuriant forests of oak and fir.

Where the rock is less destructible, its upper outline is cap-shaped, notched, and craggy, and its base encumbered with barren and ruinous piles of slaty fragments.

As examples of trachytic hummocks may be quoted, the great plateau de Bozat; that of Le Puy Gros, reaching to the village of La Queille; and that which is now divided by weathering, into the different hills called Puys de L'angle, Hautechaux, La Tache, Poulet, Baladou, L'Aiguillier, and Pessade, (Mont Dor.) That of Bort and Montagnes de Salers in the Cantal, &c.

As trachytic or phonolitic sheets or currents, those of the Plateaux de L'angle, de Rigolet, and de Chambon, in the Mont Dor; those which extend to the town of Aurillac, in the Cantal, and rest there on fresh-water marles; those of the Solfatara, and del Arso in Ischia; finally, those which constitute the substratum of Procida, Lipari, Le Saline, Volcano, Ustica, &c.

It must be recollected that the quantity and escaping force of the vapours evolved from a lava in ebullition varies directly, and the quantity of lava protruded en masse by its internal intumescence, inversely, with its fluidity. Consequently, there is nothing but what we should expect à priori in the usually stupendous volume of those trachytic masses (as in the Andes, &c.) which from the figure they have assumed, and the nature of their mineral composition, are proved to have possessed an exceedingly imperfect fluidity; while for the same reason we must anticipate that the quantity of fragmentary matters or conglomerate rocks that accompanied their production must be inversely proportionate to their bulk. It appears, indeed, far from improbable, that the protrusion of some trachytic masses whose fluidity was to the last degree imperfect, and particularly when from a sub-aqueous vent, where the density of the outer medium must oppose so great an obstacle to the absolute escape of the confined vapour, may have been wholly unaccompanied by any aeriform explosions, and consequently by the contemporary formation of any conglomerate.

It results from these observations, that there is no reason to deny the trachytic rocks to have universally issued from the earth as *lavas*, or in a liquefied state, whether produced by sub-aerial or sub-aqueous volcanic vents; that on the contrary they appear to have been elevated by the same causes and in the same manner as basalt, and that the remarkable comparative bulk of the masses in which the former have disposed themselves is simply owing to their having generally possessed, by reason of their low specific gravity, a very inferior degree of fluidity to that of the augitic lavas, which, when produced with an equally large grain, and under the same external circumstances, have always spread much further, and were therefore evidently more fluid.

It is thus that considerations, solely based upon the characteristic differences generally observable in the disposition of lava rocks of different mineral character on the earth's surface, lead of necessity to the identical conclusion at which we formerly arrived by a very different chain of proofs; viz. that under the same external circumstances, and with a similarity in the size and arrangement of their crystalline grain, the fluidity of a lava increases directly with its specific gravity, that is with the proportion of the heavier minerals in its composition.

The strongest confirmation is thus obtained of the correctness of the data from which these conclusions were drawn, which include our hypothesis on the crystalline nature of the fluidity of lava.

§. 5. After this unavoidable digression we revert to the consideration of the mode of procedure of lavas in general when poured out on the earth's surface.

If open fissures exist in the ground over which a current of lava is led to flow, part of its substance will enter and occupy them to a greater or less depth in proportion to its fluidity. The matter thus introduced between the cheeks of the fissure, assumes the form of *veins* or *dykes*, which are often with difficulty to be distinguished afterwards from the similarly disposed masses, already mentioned in a former chapter (Chap. ii. §. 15), which are produced in a somewhat different manner, viz. by forcible *injection* from below; whereas these are filled by *intrusion* from above. Every volcanic district presents examples of dykes filled in this manner, and appearing to branch out like roots from some overlying mass of rock into the supporting strata. The basaltic dykes of the chalk formation in the North of Ireland are probably in great part of this class; though many, and particularly those which appear to have affected the nature of the rocks forming the sides of the vein to some distance, were evidently the vents through which the lava rose, and was protruded to the surface. The friction produced by the resistance of the sides of the fissure to the lava entering into it in either mode, must augment the comminution of its grain. We should therefore expect to find the substance of these veins to be somewhat finer in grain and closer in texture than the overlying mass with which they communicate,—and again, that the lateral parts should be more comminuted than the centre. This observation has indeed been often made, though as yet no explanation has been offered of the fact.

When lava has once been poured forth on the earth's surface, it is urged forwards by the influence of its own fluidity, independently of the force of expulsion, so that a current continues to advance slowly for a great length of time after it has ceased to receive any supply from the volcanic aperture, and indeed after the final obstruction of this vent, and the termination of the eruption. This effect of the weight of the mass will be the greater in proportion to the slowness with which the lava yields to consolidation, its specific gravity, and the inclination of the surface on which it flows.

Thus, in the year 1819, on the flanks of Ætna, I observed a current still slowly progressing, at the rate of about a yard per day, which had been emitted nine months before; and other currents of the same volcano are cited by Ferrara and Dolomieu, as still moving on *ten* years after their emission.

As the still liquid interior of the current escapes outwardly in this manner, the upper crust must generally subside in proportion. Sometimes however the subsidence is prevented by the solidity of its structure, and it remains suspended as a sort of roof over the vacuity left by the lateral expansion of the inferior liquid. In this manner it is probable many of the caverns were formed which so frequently occur in volcanic rocks of this nature; particularly in Iceland and the island of San Miguel.

According to M. Bory de St. Vincent, the volcano of Bourbon has poured forth at various times from the lips of its crater numerous diminutive lava currents of extreme liquidity, the interior of which escaping by the influence of its own gravity and the great inclination of the surface of the cone on which they rest, has left a sort of hollow gutter, arched over by a thin and brittle crust, that frequently yields to the weight of any person stepping upon it, so as to occasion dangerous falls. As the lava retires gradually from beneath these arched crusts, pseudo-stalactitic projections, like icicles, form upon its inner surface, which are coated with a glazing or glassy varnish by the fusion of their superficies, occasioned probably by the same process as in the smaller vesicular cavities of fine-grained lavas.* (See Chap. v. §. 5.)

We have already noticed that, owing to the extremely im-

---

* See the plates of such caverns in Bory de St. Vincent's Voyage aux isles d'Afrique; and their description in Mackenzie's Iceland, and Webster's San Miguel.

perfect liquidity of lava, and the sluggishness with which it consequently moves, the slightest obstacle must have a very considerable effect in retarding its progress. And this supposition is entirely confirmed by observation. A bush, a tree, a wall, a large stone, have often been seen to check the advance of a current to an extent quite unequal to the resistance they can be conceived to oppose to the weight and pressure of the lava pouring upon them; and only to be accounted for by the idea of its semisolid condition, and the extremely low degree of mobility possessed by its particles, which the increased compression resulting from the least impediment suffices to destroy to a certain distance from the opposing surface.

Even the resistance offered by the minor asperities of the ground over which a stream of lava extends itself, destroys or lessens the liquidity of this substance to a certain distance upwards, and hence the rotatory motion with which a current appears to advance. The lower stratum being arrested by the resistance of the ground, the upper or central part protrudes itself, and being unsupported from below, falls to the ground; becomes stationary there; and is in turn covered by a mass of more liquid lava which rolls or swells over it from above. In this manner the scoriform crust that forms upon the surface of lava immediately on its exposure, is continually brought to the lower part of the current; and it is seen from hence why rocks of this origin generally rest upon a thin stratum of fragmentary scoriæ. A current which I had the opportunity of observing on Ætna in 1819, and which was advancing down a considerable slope at the rate of about a yard an hour, had all the appearance of a huge heap of rough and large cinders, rolling over and over upon itself by the effect of an extremely slow propulsion from behind. This motion was accompanied by a crackling noise, occasioned by the contraction of the crust as it solidified, and the friction of its scoriform cakes against one another; and on the whole was fitted to produce any other idea than that of fluidity.* Yet within the crevices of this sluggish mass a dull red heat might be still seen by night, and a considerable quantity of issuing vapour was visible by day.

Whenever the advance of a lava is checked by any material

* The crackling noise and rotatory motion of this current, strongly recalled the rythm of the famous Homeric line,

Αυταρ επειτα πεδον δε κυλινδετο λαος αναιδης.

when repeated very slowly.

obstacle, it accumulates upon itself, rising in height until it is able either to surmount and cascade over the obstacle, or to turn round it by a lateral deviation. The extreme difficulty with which any such diversion from the straight line of its course appears to be effectuated, is very remarkable, but by no means surprising, since this circumstance is common to all liquids of great viscosity or consistence, which when urged forwards down an inclined plane by their gravitating force, move, as it were, *en masse;* the component particles retaining almost completely their relative positions, without rolling over one another in that free and voluble manner which characterises the motion of more perfectly liquid bodies.

In fact, we must represent to ourselves the mode in which the crystalline particles of lava move amongst one another, rather as a sliding or slipping of their plane surfaces over each other, facilitated by the intervention of the elastic fluid, than as the rotatory movement which actuates the globular molecules of most other liquids. When, in a liquid of this character, a rectilinear motion has once commenced, the particles will arrange themselves in such a manner as is most favourable to their easy progression in that direction, but this very arrangement will increase the difficulty opposed to any change in the direction of the movement.

If, for example, we take the clinkstone lavas, which we find composed of crystals generally having one or two dimensions greatly superior to the third; (in the trachytic clinkstones the crystals of felspar being thin tabular scales, or long, narrow, and shallow rhomboids; and the occasional crystals of augite or hornblende which occur in them being generally acicular;) it is obvious that the communication of liquidity and motion to such a mass will cause the crystals to arrange themselves in such a manner as to have their longest axis in the direction of the motion. This effect will be necessarily produced by the mutual friction of the particles against each other, to which those that lie transversely, across the direction of the motion, are most exposed, and are consequently forced to accommodate themselves to that direction.

After such an arrangement, a very small quantity of vapour interposed between the parallel surfaces of the proximate crystals will suffice to allow of their gliding past one another in the direction of their longest dimensions; while motion in any other direction transverse to their parallel planes is rendered far more difficult, and requires a proportionately greater force to produce it.

In the same manner, in every species of lava when once set in motion, the crystals will take such an arrangement as offers least impediment to their moving amongst one another in the direction impressed on the lava, their longest axes assuming a coincidence with that direction: so that should the direction of the motion be subsequently changed by any external force, the actual arrangement of the crystalline particles being unfavourable to their new motion, will offer considerable resistance to it, and check it more or less, until a new arrangement is impressed on the particles consonant to the new impulse. Until this is accomplished the quantity of vapour which communicated great mobility in the former direction may be insufficient to permit it to take place at all in the latter. Hence it is seen how a lava often possesses great fluidity in one direction, and little or none in another; this difference being always proportionate to the average difference between the longest and shortest axis of its component crystals. This sliding or glissant mode of progression accounts not only for the extreme difficulty with which a stream of lava is induced to alter the direction it has once been led to assume, but also for the fact that so few of the larger crystals occurring in rocks of this origin are broken or rounded at the angles, as would have been the case supposing them to have moved freely amongst one another in a rotatory manner through the medium of their elastic vehicle.

It is however impossible, but that a considerable mutual friction of the particles must accompany the progression of every lava current, and this will be increased by the occurrence of obstacles forcing it to swerve repeatedly from a rectilinear course. It is a remark which I have frequently made, that the grain of a lava rock is generally finer at the extremity than at the source of the current, particularly if it has flowed to a considerable distance. Analogous to this fact, is one mentioned by De Buch, with regard to the lava currents of the island of Lancerote, produced by the eruption of 1730; which contain near their origin innumerable knots of olivin, many of them as large as a man's head. These dwindle away in size towards the extremity of the currents, where scarcely any are visible; the olivin being there uniformly mixed up with the other crystalline grains of the lava.—(De Buch on Lancerote.)

It frequently must happen that some of the superficial parts of a lava current, or those which are consolidated by the resistance opposed by some obstacle to the motion of the current, are broken up by the renewal or continuance of its motion, and their fragments enveloped in the still liquid lava, and rounded by the friction of its particles.

Hence the *brecciated* aspect which is occasionally met with in rocks of this nature, and the imperceptible manner in which the fragments often appear to graduate into the base.—(See Memoir on Ponza Isles, Geological Transactions.)

Many other appearances presented by masses of lava after their solidification are also obviously accounted for by these considerations: thus in the clinkstones or slaty lava-rocks, of which the characteristic feature is the parallelism of their component crystals, the directions of the longest axis of these crystals will, I believe, be found universally coincident with the direction of the motion of the mass in a liquefied state; and, as might be expected à priori, it will, I think, be generally recognised from observation, that the lavas of this character possessed a much higher fluidity in proportion to the size of their component crystals than ordinary trachytes and basalts, in which the crystals have their different dimensions more nearly equal, and which are consequently irregularly aggregated.*

The circumstances by which this peculiarity of crystalline form was probably occasioned will be adverted to at a future opportunity. (See Chap. v. §. 2.)

In the same manner the zoned or ribboned structure of the trachytes of Ponza, of the trachytic porphyries and pearlstones of Hungary, Lipari, and the Andes, in which layers of different texture repeatedly alternate together, or a base of one texture encloses elongated lenticular masses of another, is evidently owing to the irregular parts of unequal texture, of which the lava consisted, having been *drawn out* or elongated during its fluid state in the direction of its motion.†

In these, and I believe in all lava rocks, those imbedded crystals of felspar, augite, or hornblende, which have any one dimension much longer than the others are found to have their longest axis generally disposed in the direction of the movement.

Owing to its sluggish and peculiar motion, a lava current

---

* The great clinkstone current from the Mont Mezin in the Department de la Haute Loire, already quoted, offers a remarkable instance in corroboration of this assertion. In the Mont Dor and Cantal, the same fact is observable, viz. that the currents of clinkstone have generally spread to a greater distance from the centre of eruption than those of granitoidal or amorphous trachyte.

† Memoir on the Ponza Isles. Geol. Trans. 1825.

increases considerably in depth and bulk wherever any impediments retards its progress. Thus for example, those which, in the Provinces of Auvergne and the Vivarais, have occupied the granitic channels of mountain torrents, flowing to a distance of five, ten, or fifteen miles from their source, may be observed to swell out in width and thickness wherever the sinuosities of the valley produced a deviation in their direction; so that every concave elbow of these gorges is filled by a bulky mass of basalt, while the intervening parts of the current present comparatively narrow and shallow strips.

These latter portions have in some instances been subsequently, in part or altogether, worn away by the erosive action of the torrent; and the huge patches of basalt, which are thus left alone and unconnected in the bosoms of the alternate angles of the valley, have a strange and a striking appearance, heightened in general by the excessive regularity of their columnar configuration. (Vals d'Antraigues, de Monpèzat, &c.)

If the obstacle be of importance, and so extensive in a horizontal direction that the lava cannot turn, but is obliged to surmount it, the accumulation necessary for this purpose will often attain a much higher level in front of the obstacle than the surface of the current at some distance behind it, so that the stream has the appearance of having flowed up hill. The cause of this is that the hardened crust acts the part of a covered canal, in which the liquid rises to its level like water in a pipe.

When a lava current has met in its course with a flat and extensive surface perpendicularly opposed to its direction, such as the wall of a house, &c., it has been observed to stop, as if by magic, at the distance of 10 or 12 inches, without coming into actual contact with the obstacle. This extraordinary fact, denied by those who have not witnessed it, perhaps only because it appears at first inexplicable, is perfectly reconcileable to our hypothesis as to the nature of the fluidity of the substance.

A vast quantity of elastic vapours, at a high degree of tension, escape from every fresh surface of lava successively disclosed as it moves onward; but as the lava approaches close to a resisting surface of considerable extent, their freedom of escape must be prevented in proportion, and the reaction of the elastic fluids evolved from the lava, and filling the narrow space or crevice intervening between its surface and that of the opposing wall, must create a resistance sufficient sooner or later to check its further advance.

The lava is then consolidated to some distance back by the pressure of the current behind. If the momentum of this mass be considerable, the wall may give way; but should the motion of the lava be slow, and its fluidity very imperfect, it may rise without overthrowing or even touching the wall, until it is sufficiently elvated to cascade over it, or deviate in a lateral direction. This curious observation has been repeatedly made during all the destructive eruptions of Vesuvius, and there seems no reason to doubt its correctness.

When there has existed a wooden door in the wall, it has been observed, that the heat radiated from the lava, after some time, sets fire to it, and when thoroughly consumed, the lava enters the aperture thus produced, but continues to respect the wall on either side. The reason of this is obvious; for, since the lava is only checked by the reaction of the elastic fluids it evolves from any resisting surface to which it approaches, on those points where no such surface exists, or, having existed, has given way,—the lava experiencing no obstacle advances in the usual manner.

The dry grass or shrubs which a current meets with are frequently set on fire, and the flames produced in this manner and seen from a distance, have no doubt often been mistaken for the volcanic phenomena.

When trees are rapidly enveloped in lava, the upper parts alone blaze and are reduced to ashes, the trunk is merely carbonized, and if subsequently removed by meteoric injury, may leave its impression in a hollow-cylindrical tube within the solid rock. Such moulds occur very commonly in those currents of the Isle of Bourbon, which have extended their ravages through forests of palm trees, (Bory de St. Vincent, Voyage.) and one of these is to be seen in the cabinet of the Jardin des Plantes at Paris.

The fragments of pre-existing rocks accidentally enveloped by lava in its advance, are observed to be variously affected by the heat. In general they are partially fused on the exterior, and incorporated so closely with the surrounding substance that it is difficult to trace their outline with clearness. Fragments of limestone still preserve their carbonic acid, but have often acquired a crystalline grain, probably from having been partially fused under pressure, as in the experiments of Sir James Hall. Near Aurillac in the Cantal, I have found fresh-water shells of this character incorporated in the lowest stratum of a basaltic current, which appears to have flowed from the heights of the mountain into the lake that existed at its foot. Fragments of sandstone enveloped in this manner are usually hardened, and clay is frequently converted into a substance resembling jasper (porcelain jasper) —in other instances into *Tripoli.*

Owing to the immediate solidification of lava flowing in the open air, as soon as it impinges on any obstacle, it cannot of course radiate sufficient heat to produce any important change in the substance of these impediments. The numerous lava currents which in the Auvergne have flowed to a great distance over both granite and limestone, have rarely effected any alteration except on the mere surfaces of those rocks with which they have come in contact. When Torre del Greco, however, was overflowed by the lava of 1794, some curious circumstances were observed as to the effect of its caloric on the metallic and earthy substances with which it came in contact. Both were in part volatilized, and deposited as crystalline sublimations on neighbouring points. The component metals of some alloys crystallized separately.*

If a current of lava flows over marshy ground, the sudden vaporization of the humidity confined below must occasion explosive discharges, that will disturb, and occasionally tear their way through the superincumbent mass, scattering its fragments into the air, and thus creating either rude hollows or scoriform protuberances on the surface.

When a current enters the sea, or any other body of water, a similar effect may be in part produced, but these explosions are of no very extraordinary violence. It has been supposed that on such an occasion a violent combat must ensue between the two elements, and many very poetical images have been employed to heighten the colouring of the picture; but the fact appears to be quite otherwise. A certain quantity of the water that approaches nearest to the incandescent lava, as it is protruded from the extremity of the current, will be reduced to vapour; but the superficial consolidation that instantly ensues will prevent any further contact. The fissures that open in this crust as it solidifies, emit torrents of vapour the instant they are formed, which must impede the entrance of the surrounding water till their sides are consolidated and comparatively cooled.

From these considerations, as well as from all the direct observations that have been made on the subject,† we must believe that a current of lava advancing below water con-

* See Brieslak, Voyage dans la Campanie, tom. i. page 278.
† See Monticelli, Brieslak, &c. passim.

ducts itself in much the same manner as on dry land. Its lateral progress will however be necessarily much slower, since its liquidity will be diminished by the superior density and weight of the surrounding medium acting in opposition to the expansion of the vapour it contains. At the same time the density of the water will proportionately impede the escape of the steam, and the superficial consolidation will rather take place by the cooling and condensation of this va-vapour by contact with the cold water, than by its absolute escape.

The caloric abstracted from the lava, both by actual contact, and by the condensation of its heated vapour, soon communicates a proportionately high temperature to the water in its vicinity, which on such occasions is observed to be heated, discoloured, and rendered turbid to some distance. Fish have been often found killed, in very considerable numbers, by this sudden change of temperature in their native element.* It is far from improbable that the Icthyolites of Monte Bolca were destroyed by an occurrence of this nature; since the fissile limestone, in which they are imbedded, is immediately capped by a bed of basalt and calcareous peperino.

The latter rock proves the eruption that produced the basalt to have been submarine, and the remarkable attitudes of some of the fish bear evidence to the extremely sudden nature of the catastrophe that at once destroyed, and buried them in the soft calcareous sediment, which was at that time habitually deposited at the bottom of the sea.

---

* During the great eruptions of Lanzerote, one of the Canaries, which continued unceasingly, and with intense violence, during the years 1730—1736, immense multitudes of fish are said to have been cast ashore dead. The same in Iceland, 1783; in the eruption of Vesuvius, which destroyed Torre del Greco; at Stromboli, &c.

CHAP. V.

Consolidation of Lava.

§. 1. The consolidation of any mass of lava is effected, either, by the *condensation* of all the aqueous vapour which it contains, and to which alone its liquidity is owing: or, by the *escape* of a part of this vapour, and the condensation of the remainder.

§. 2. We have seen that wherever the expansive process has been so far developed as to *disintegrate* any of the component crystals of the heated lava, the vapour formed in it exists in two different conditions—part being *fixed*, i. e. still imprisoned between the crystalline laminæ, where it was generated from the water mechanically contained there, and creating by its elasticity a sort of repulsion between these laminæ, without altering the relative positions of their axes;—the remainder *free*, having liberated itself from the points on which it was vaporized, by displacing the confining laminæ, and thus more or less disintegrating the whole crystal in which it was formed; that is, breaking it up into smaller crystals unconformably disposed towards one another, between which the liberated vapour remains in irregular parcels.

The condensation of the vapour in either of these conditions, is occasioned by an increase of pressure, or a diminution of temperature. In the first instance, the forces which tend to bring the particles together, being augmented so as to acquire the ascendancy over the expansive force of the vapour;—in the second, the elasticity of the vapour being diminished, so as to give the ascendancy to these unaugmented forces, which consist of the weight of the overlying lava;—the resistance of the remaining general expansive force of the liquid, from the obstacles to its full development; and the attractive forces of the crystalline molecules.

By the gradual condensation of the *fixed* vapour, the laminæ of those crystals which the process of expansion has only *disaggregated*, are again brought together, and ultimately by the complete condensation of the intervening vapour, without any derangement in the position of their axes, are reaggregated into whole crystals, of precisely the same form as before their disaggregation.

The condensation of the *free* vapour, in the same manner allows the approach of the disintegrated crystals, which its elasticity kept asunder, but the poles of these crystals no longer correspond, and their proximate plane surfaces are no longer parallel: consequently their reaggregation, in their actual relative positions, must leave a number of irregular interstices between them, and produce a confusedly crystalline, (granitoidal) and more or less porous rock. But by this forcible reunion, in so irregular a manner, it must be expected that many angles of the crystals will be broken up, and forced into the interstices of the larger crystal: by which the porosity of the resulting rock will be reduced, and in proportion to the force with which the reaggregation is produced. This must be greater when the condensation of the interstitial fluid takes place by increase of pressure, than when it is occasioned only by the diminution of temperature; hence the compactness of the lava rock will be greater in the first case than in the last.

Again, should the condensation of the intermediate elastic fluid proceed with extreme slowness, so that, while the diminution of its elasticity brings the particles by slow degrees within the sphere of their reciprocal attractive forces, the remaining elasticity leaves a sufficient mobility to permit of the reversion of their poles in obedience to these forces: many of the particles may be enabled to unite into larger crystals; and thus more or less of a partial recrystallization occurs.

It is easily perceived how this result will be facilitated, and therefore how the number, size, and perfect form of the crystals reproduced from the comminuted and disintegrated crystalline particles, will be augmented, in proportion to the slowness with which the interstitial fluid is condensed: whether this condensation take place by increase of pressure, or diminution of temperature. And this reasoning agrees entirely with the results of experiment, and is particularly confirmed by those of Sir James Hall and Mr. Watt, on the crystallization of rocks, from fusion, by slow refrigeration.

It is obvious that the particles of such an imperfectly liquid mass, will experience a much greater resistance to their motion in the direction of the pressure to which they are subjected, than in any other; and that the resistance will be least to movements taking place in directions at right angles to that of the pressure. Consequently the process of crystallization will be soonest impeded in the direction of the pressure; and the crystals formed will have their shortest axes in that line, and their longest in the plane to which it is perpendicular; and the whole mass will have more or less of a lamellar or foliated structure, such as is observable in the clinkstones, gneiss, &c.; the laminæ taking an arrangement perpendicular to the direction in which the opposite forces of expansion and repression are exerted. It is highly probable that the lamellar and foliated lava rocks owe this peculiarity of structure to their having more or less completely recrystallized within the volcanic vent, during the continued predominance of the repressive force over the expansive efforts of the more heated lava beneath.

§. 3. But of the vapour contained in a mass of lava undergoing the process of consolidation, *that* portion *only* will be condensed which cannot effect its absolute *escape*.

In those masses which are consolidated, whether by pressure or loss of caloric, beneath an overlying crust of rocks, or confined within a narrow fissure, as in the case of dykes, little or no vapour can effect its escape in any manner, and the whole will be condensed.

The condensation in either of these cases will be necessarily very slow, and the resulting rock proportionately compact.

In those masses of lava, on the contrary, which are consolidated by exposure to a fluid medium of great rarity compared to the elastic fluid which acts as its vehicle, a certain proportion of this fluid may be enabled to effect its escape, in one or both of the following methods, viz.

I. By the ascent of the vapour in obedience to that elevating force which consists of the difference of its specific gravity to that of the surrounding liquid mass.

This rise will be accompanied by the agglomeration of the neighbouring parcels, in consequence of their mutual affinity, as their motion brings them within the sphere of exertion of this force. The more perfect the liquidity of the lava, that is, the finer its grain, the more nearly will these compound parcels, or *bubbles*, approach to the *spherical* form, the pressure being the more completely equalized on all their sides.

If the lava which contains them moves by the influence of its own gravitation in any direction, these bubbles will be drawn out, or *elongated*, in the direction of the motion. If, on the contrary, the motion already acquired is impeded by any obstacle, the bubbles will take the form of flattened spheriods, having their shorter axis in the direction of the impulse. These will be the necessary results of the pressure being *diminished* in the first case, and *increased* in the last, in the direction of the movement. For the same reason, if the vertical pressure upon any of these bubbles is at any time diminished, their expansive force, taking effect in that direction, will occasion a vertical elongation of the globule.*

As these bubbles traverse the lava, they are continually enlarged by the accession of fresh parcels of vapour, which they approach in their ascent; consequently their volume will be proportioned to the space they have traversed, and this will depend in turn on the specific gravity and liquidity, or the fluidity of the lava; which alone determines the velocity of their ascent, as well as the proportion of the whole quantity of vapour contained in the lava which effects its escape in this manner, to that which remains behind. (The latter quantity is always either exactly, or, less than, what is necessary to communicate a liquidity sufficient to permit this union of the several parcels in their ascent). Of this latter quantity a certain portion will subsequently escape in another and very different manner, which forms the second mode of escape.

II. While the rise of vapour in bubbles requires a high degree of fluidity in the superior mass of lava this second mode of escape is facilitated—on the contrary by the lowest degree of fluidity amounting almost, or entirely, to solidification. It takes place by the influence of the expansive force of the steam, which urges it in every direction where an avenue is afforded for its passage, either through the pores that are left in the external parts of the lava as it solidifies, or the fissures that are formed by its contraction during the same process.

The partial escape of the vapour in this manner, by diminishing the elastic spring of what remains, acts in precisely the

* Examples of these various forms are common in fined-grained lavas. The last is particularly exemplified in the obsidian streams of Lipari and Iceland.

same manner as its condensation, in producing the reaggregation of the disintegrated crystals; but their reunion, in this case, usually occurs in too rapid and tumultuous a manner for the process of recrystallization to take place, and will therefore be irregular and nonconformable, (or granitoidal,) and more or less porous, little or no change taking place in the position of their poles.

The porosity of the rock, that is, the space occupied by the interstices of these crystalline particles, when reaggregated by the partial escape of the interstitial fluid, will be determined by,

1. The arrangement of the particles, whether more or less conformable; the compactness of the rock being directly proportioned to the parallelism of the proximate crystalline surfaces.

2. The average size of the crystalline particles. If these are uniformly minute, the rock must be proportionately compact; the small particles, by reason of their mobility, fitting closely into one another, so as to leave few or no interstices.

If the grain is uniformly coarse, the interstices of the particles will be large, and the rock proportionately porous.

If the grain is unequal, consisting of large crystals merged in a basis composed of much more minute particles, which acts as their vehicle, the compactness of the basis will be proportioned to the comminution of its component particles.

The compactness of the imbedded crystals will also vary directly with that of their basis: since in proportion to the liquidity of this basis (that is the fineness of its grain) will be the degree of pressure to which they are subjected previous to the consolidation of the surrounding mass. After this has taken place, but not till then, the vapour fixed between the plates of the larger crystals may also in part effect its escape, by percolation through the interstices of the enveloping solid, leaving the crystals themselves more or less cracked, fibrous, and disintegrated; no pressure acting upon them to bring their laminæ again into complete contact. Hence the striated and cracked appearance of so many of the large imbedded crystals of the trap-rocks.

The heat radiated from the enclosures of these crystals, after their contained vapour has partly escaped, may have produced that partial fusion of their fibres which is occasionally visible, as in the trachytes of the Mont Dor, &c.

This solidifying process must begin from without, and spread inwardly, with a rapidity, and to a depth, proportioned to the porosity of the rock resulting from the external consolidation of the lava; which, we have seen, varies directly with the size of its crystalline particles, and the irregularity of their mutual arrangement.

The depth to which the fissures penetrate that are formed by contraction as the mass solidifies, must obviously coincide with the depth of this process, and consequently follow the same law.

Hence we deduce as a general proposition, that

The process of consolidation effected by the exudation of the vapour through the porous interstices of the outer parts of the rock already consolidated, proceeds from the exposed surfaces inwardly, in all directions, with a rapidity directly proportionate to the size and irregular arrangement of the component crystals; and that the looseness of texture, or porosity of the rock is determined by the same law.

The penetrating power of the elastic vapour, urged to escape in this manner, is probably very great; and there are few of the granitoidal, or confusedly aggregated rocks, which are not more or less permeable to it.

As a corollary of the above proposition, we observe that the compactness of a rock, supposing the size and arrangement of its crystalline molecules fixed, will vary inversely with the quantity of vapour which escaped from it by exudation. And again generally—that of the whole quantity of vapour generated in a mass of lava, the proportion that escapes, by *rising in bubbles*, will vary directly with the specific gravity of the lava, ceteris paribus;—that which escapes *by exudation*, with the size and irregular position of the crystalline particles;—that which remains enclosed, and is condensed by pressure, varies inversely with the specific gravity, and irregularity, and directly with the size of the crystalline particles: and therefore will be inversely proportionate to the specific gravity, and comminution, and directly to the conformable arrangement of the compound crystals.

§. 4. From the foregoing propositions we deduce the following considerations, as to the conduct of a mass of lava protruded on the surface of the earth.

I. If its crystalline particles average an extreme degree of comminution, and at the same time the specific gravity is at the mininum, the superficial crust that instantaneously congeals upon exposure will be thin, compact, close-grained, and more or less glassy. The vapours of the interior will unite perhaps into bubbles, owing to its extreme liquidity, but will not be enabled to escape by percolation through this crust, in consequence of its close grain: nor to break it up by the force of their ascent, since, by reason of the low specific gravity, this force is at its minimum. Consequently the lava will intumesce considerably; and the reaction of the confined vapours from the crust will create a pressure, by which their further expansion is prevented.* Such a mass will retain its heat very long, cooling down only by slowly transmitting its caloric (by radiation) to the surrounding bodies, and subsiding gradually as its contained vapours are condensed by the diminution of temperature.

Its fluidity would be considerable, the extreme comminution of its grain compensating in part for its low specific gravity; and the vapour bubbles that form within it will therefore be almost perfectly spherical.

As it flows, these bubbles will be elongated in the direction of the motion. At the same time those parts in which the vapours had first collected into bubbles, will be likewise drawn out, in the same direction, into lenticular masses, or zones, alternating with the intermediate or enveloping parts, in which the vapours remain fixed—these masses will be still further flattened, by the slow subsidence of the mass. The bubbles occasion a certain degree of resistance to the

* The same cause that produces the vesicles in our loaves while baking, occasions those of obsidian, pomice, scoriæ, and cellular lavas in general, viz. the generation of elastic vapour in a substance of imperfect liquidity, which is prevented from escaping by the induration of a superficial crust on the outside of the mass.

The mixture of butter in the dough destroys its cohesiveness when exposed to heat—prevents the formation of an impervious crust—renders the substance granular, and allows the escape of the vapour generated within it, by exudation.

Hence the different texture of *cakes*, which may be compared to the larger grained and porous trachytes, and bread, which resembles close-grained and compact, or fine basalt, or obsidian currents. May I be excused this homely but apt illustration?

motion of the enclosing liquid, and therefore those parts in which none exist, will bear the appearance of having moved faster than those in which they have formed.

In the same manner, whatever large crystals or fragments of unequal dimensions are contained in the lava, will be arranged by the friction of the moving mass, so as to have their longest axes parallel to the direction of the movement.

Such are precisely the appearances of the obsidian currents of Lipari, Volcano, Iceland, Teneriffe, Bourbon, and the Andes. Whatever crystals appear imbedded in these rocks, rendering them porphyritic, (pitchstone porphyries, &c.) are clearly not of subsequent formation, but existed in them prior to the emission of the lavas, as is proved by their broken, and half-fused state. The surface of these currents is usually extremely vesicular and harsh, or in the state of pomice, which is merely vesicular, filamentous, and highly disintegrated trachyte.

Whatever ejections of scoriæ accompany the production of such lavas, they will consist of pomice, &c.

But it has been observed above, that the solidification of any lava by the slow condensation of its contained vapour, whether by gradual loss of temperature, or increase of pressure, creates circumstances favorable to the exercise of the forces of affinity among the component particles, and we should therefore expect some appearances of this process having taken place in the case under consideration.

Such an arrangement seems in fact to have occasioned the globular concretionary structure which characterises *pearlstone*; and which is seen also in the fine-grained trachytes, and trachytic porphyries, of Ponza, and Hungary. It would appear that the formation of these globular concretions requires considerable freedom of motion amongst the particles, and is therefore favored by an excessive fineness of grain, and a low pressure: their structure being more perfect and frequent, according to these circumstances. When most complete, the globules are radiated from the centre to the circumference, as well as divided into concentric coats.

This process is evidently a species of crystallization, the finer crystalline particles reuniting according to the laws of their affinity around some nucleus, which acts as a fulcrum for the crystals to turn upon. This nucleus is generally a minute pre-existent crystal, at other times a small solid grain of the glassy base.

This partial recrystallization occasions a fine crystalline grain in the globules: which, when they have formed from a lava disintegrated, almost or entirely, to its integrant molecules, (that is, in *complete fusion*) produces a strong contrast between their dull or lithoidal texture, and that of the glassy mass which envelopes them. Hence they have been called vitro-lithoidal globules, or *crystallites*.*

These globular concretions seem generally confined to the felspathic lavas. This is obviously owing to the circumstance that this structure, requiring a low pressure, can only take place when the specific gravity of the lava is low. Occasionally, however, it is observable, in a less perfect degree, amongst the *greystone* lavas, when the disintegration has proceeded sufficiently far; giving rise to what has been called the *variolitic* structure. This is frequently rendered visible only by a commencement of decomposition, which, acting first upon the globular concretions, in which the felspar is in excess, and reducing them to a sort of kaolin, gives to these a much lighter tint than the more augitic base. The globular concretionary structure prevailing generally amongst the felspathic lavas, felspar is of course usually the mineral which has agglomerated in this manner by the force of its affinity. This is universally the case in the pearlstones, trachytic porphyries, &c.; but where augite bore any proportion as an element to felspar, and the grain of the lava was sufficiently comminuted, the former mineral has occasionally concreted into globules, more or less imperfect, as in the spotted clinkstones (phonolites tachetées) of Sanadoire, Mt. Dor, Mercœur, and Mezen; of Helsburg in the Cobourg territory; of the Mittel Gebirge, and other parts of Germany.

In the former case, the spots are light on a darker ground; in the latter they are dark, and greenish, or grey, on a lighter ground.

These globular concretions have the same effect which was attributed above to the vapour bubbles, of diminishing the

---

* We know that when common glass is allowed to cool very slowly from fusion, a similar process takes place, concretionary crystalline globules are formed in it, which have been called *vitrites*.

This species of crystallization is the same which gives rise to the globular pyrites of chalk and clays which usually collect around some pre-existent nucleus.

The globular granite of Corsica is a similar instance in a more coarsely crystalline rock.

The flints themselves of chalk are imperfect agglomerations of the same kind, all tending more or less to the globular form, when the force of gravity and pressure did not interfere too powerfully, during the imperfect liquidity of the mass.

liquidity of the lava: and hence, if after this process has taken place, it should continue in motion, or a fresh impulse should be communicated to the mass, the parts in which this structure occurs will move more slowly than the rest, and, as in the former case, these parts will be drawn out into long streaky or lenticular masses, enclosed in and alternating with zones of a finer grain, in which no concretions occur.*

If a diminution of pressure accompanies the motion of the lava, the globular concretions will be more or less disaggregated by the vaporization and expansive force of the water they contain: in the same manner as the more perfect crystals of coarser lava.

This disaggregation, together with the friction they sustain from the more fluid substance which envelopes them, will perhaps disintegrate them further, and draw them out into long parallel zones or streaks in which all traces of the globular structure will be lost, and which will distinguish themselves from the intervening matter only by their coarser crystalline texture, and purer mineral composition, producing a ribboned and marbled aspect in the mass, and occasionally a schistose, or lamellar structure.

The different rates of fluidity of these different parts occasioned by the variations in the size of their grain, will often give rise to curvatures, and irregular contortions in the parallel zones—these are sometimes of the most complicated character.

The trachytes of Ponza offer an example of lava-rocks, in which all the several stages of this process may be distinctly followed. (See a Memoir on Ponza Isles, Geol. Trans.)

II. If we suppose a higher specific gravity to be united to extreme fineness of crystalline grain, the fluidity of the lava will be proportionately increased; and with this, the extent of its lateral spread, under the same circumstances of slope and productive ratio.

The bubbles of steam formed in the interior will also rise, with a force, and in a quantity, proportioned to the specific gravity: tossing up the liquid as they burst from its surface; and since the rapidity of superficial congelation varies with the liquidity of the lava, (i. e. its fineness and irregularity of grain, *ceteris paribus*) the fresh surfaces, exposed by this tu-

* Instances of this are common to all the pearlstones of Hungary, the Andes, Kamskatka, &c.

multuous agitation, will be instantly consolidated; and the external parts of a lava current of this character, will present the utmost roughness and irregularity, bristling with glassy, ragged and frothy asperities, so as to suggest the idea of a stormy sea congealed at the instant of its wildest agitation.

Those bubbles which, by the suddenness of the superficial congelation, are imprisoned within the lava, will remain near its surface, forming vesicular cavities, the size, number, and sphericity of which, will be proportioned to the specific gravity, and fineness of grain.

Where these are at their maximum, large cavernous blisters may be formed, such as occur in the lavas of Iceland, Bourbon, Lanzarote, &c. Some of them will burst, and allow the escape of the steam through the aperture, previous to the consolidation of the entire surface, but after the process had sufficiently taken place to prevent the falling in of the sides when the elastic fluid has escaped. The walls of these cavities, both on a small and large scale, exhibit in general the traces of a certain degree of superficial fusion, in a glassy varnish which appears to be more or less remarkable, in proportion to the fineness of the grain, and the size of the vesicles. The same kind of glazing appears on the exposed surface of the lava when rapidly consolidated.

It may be accounted for perhaps by the consideration that the instantaneous vaporization and expansion of all the water contained in the superficial pellicle of lava, suddenly lowers its temperature at the moment of its solidification. In the next moment, therefore, the equilibrium will be restored, by the rapid transmission of caloric from the intensely heated lava immediately beneath, which must act upon the superficial pellicle, in the same manner as an increase of temperature produced in any other mode, and occasion its partial fusion; which will be the more easily accomplished in proportion to the fineness of its grain.

When the vesicular cavities are large, pseudo-stalactitic protuberances of lava will form upon its upper and lower surfaces, by means of the tenacity of the mass as it is forced asunder; and if the lava contain solid crystals, or knots of any substance, the weight of these will occasion them to project in lumps, or at the extremity of pendulous filaments into the interior of the blister. (As the olivin masses of the lava currents of Lanzarote, described by De Buch.)

The large hollows that are frequently formed immediately under the upper crust of a loaf, when baked in a hot oven, afford a homely but apt illustration of these natural caverns, which are produced by the operation of extremely analogous causes.

But in proportion to the quantity of vapour that escapes entirely in this manner, will be the compactness of the lower parts of the lava current; and therefore, with the same size and arrangement of grain, the compactness of the lower part of a lava current will vary directly with its specific gravity; and vice versâ, when this latter term is fixed, the compactness varies with the comminution, and regular arrangement of the crystalline particles.

On the other hand the asperity, glassiness, and harshness of the surface, will vary directly with the compactness of the interior; and since the sharper and more ragged the surface, the more brittle will it be, and the more easily must it yield to mechanical destruction, it is obvious why the upper parts of the older and compactest basaltic currents, having been long exposed to violent denudations, seldom present any remains of the scoriform and vesicular surface which they probably once possessed. The results of the play of affinities which takes place during the consolidation of such a lava will be considered in the next section.

III. When the specific gravity of a lava is high, but its grain more or less coarse, as well as irregular, its fluidity, and the lateral extension of the current is proportionately reduced, and its bulk of course increased.

The crust that in this case forms by superficial consolidation,. is more or less thick, massive, and porous; a certain proportion of the vapour escapes by percolation and through the cracks of retreat, which are wide and deep in proportion to the size of the grain.—But owing to the high specific gravity, the bubbles formed in the interior may still possess a powerful force of ascent; and will therefore, as they rise, break up the outer crust into flat, thick, and scoriform cakes or solid angular blocks. The contraction of the mass, which produces fresh horizontal fissures at right angles to the vertical ones first formed, assists in breaking up the surface; and the motion of the lava still flowing or rolling on below still further displaces the fractured crust. Hence, the surface of such lava currents will be extremely rugged and uneven, presenting rude heaps of shapeless and fissured blocks, suggesting no idea of a previous liquidity, but resembling rather the rocky debris of a shattered mountain. Such in fact is the aspect of all the coarse-grained basaltic lava currents which cover the slopes of Ætna and Vesuvius, of the cheires de Come and de Nugère in Auvergne, and of many of the lava currents of the Puys brulé of Bourbon. In the latter spot the contrast between the rude and ruined surface of the coarse-grained lava-streams; and the smooth, and, as it were, varnished, outside of the finest grained,* affords a strong confirmation of the principles on which we are proceeding.

The compactness of the interior of the current will be directly proportionate to the quantity of vapour that effects its escape in bubbles, and inversely to that which escapes by percolation, consequently it will vary directly with the specific gravity and fineness of grain.

When the comminution of the grain bears so small a ratio to the specific gravity of the lava as to prevent the ascent and absolute escape of any steam, except in bubbles from the immediate neighbourhood of the surface, the vapour below may still unite into irregular compound parcels, which will remain disseminated through the mass, more or less generally, in proportion as they have been permitted, or not, to rise from the points on which they were formed; producing the small cellular cavities of indeterminate and often angular forms, which characterise many cellular lavas (as for example those of Volvic and the Puy de Come, and that used as building-stone at Mont Dor les Bains in Auvergne; that of la Scala, at the foot of Vesuvius, &c.

IV. A still lower specific gravity joined to considerable coarseness of crystalline grain, will wholly prevent even the union of the parcels of free vapour into bubbles, and preserve them uniformly disseminated through the mass, between the disintegrated crystals, whence they slowly escape by percolation, as the mass consolidated from without inwardly; leaving *pores*, i. e. minute irregular interstices, between the crystalline particles, which unite in a confused and irregular, or granitoidal manner, as their elastic vehicle escapes. The fluidity of such a lava is extremely low: and the bulk of the currents or beds, in which it is disposed, proportionately great. Such are all the loose earthy Trachytes; the Lava Sperone of the Alban hills, the Pietra di Sorrento, and Piperno, of Naples, &c.

* Bory de St. Vincent.

Previous to the solidification of a lava of this nature, a separation of the different component minerals appears occasionally to have taken place; exactly similar to that concretionary process by which flints were formed within chalk, solid marl-balls, or fine-sandstones, within clay. Thus in the lava-rock called Piperno, of Pianura, and the Pietra di Sorrento, near Naples, as well as in the lava Sperone, and that of St. Agata, on the road to Rome, the finer particles of augitic matter have agglomerated into innumerable impure and irregular concretions, within the mass of the rock, which is principally felspathic. In the disposition and aspect of these masses, a strong confirmation is afforded of the views disclosed above, on the conduct of lavas of different qualities. Thus, while the felspathic base is of loose texture, and extremely porous, the augitic concretions are finer grained, and comparatively compact, and, from their superior specific gravity, filled with vesicular cavities, instead of irregular porous interstices. The form of these masses is either lenticular or venous; their shortest dimension being in a vertical direction, and their longest in the direction in which the mass flowed. The subsidence of the mass during its consolidation, was probably also in part the cause of the horizontal extension of these nodules. Their elongation in the direction of the currents, as well as the vesicular cavities produced by the ebullition of their aqueous particles, prove the concretionary separation to have preceded the settling of the lava in its present position.

V. When the specific gravity is equally at or near its minimum, but the size of the component crystals still coarser, that is, less disintegrated; the greater proportion of the vapour will remain fixed, and therefore will be condensed, by gradual cooling, being unable to effect its absolute escape in any manner; and consequently the resulting rock will be more compact, and freer from pores.

If the arrangement of the crystals be irregular and nonconformable, the fluidity of the lava will be at its minimum; as in the large-grained Trachytes, Syenites, Diallage rock, granite, &c. But should they be conformably arranged, the lava may possess a considerable fluidity in the direction of the longest plane surfaces of the crystals, as appears to have been the case with the trachytic clinkstones.*

* And perhaps with gneiss, and the other schistose crystalline rocks, &c.

In the former case, the lava, when protruded on the surface of the earth, may remain almost perpendicularly above the vent, and have scarcely any, or no, lateral spread. In the latter, it may assume the disposition of hummocks, sheets, or long extended currents.

VI. When the rock has been scarcely, if at all, disintegrated, being elevated, *en masse, in a solid state,* by the expansion of lava at a great depth beneath (after having been gradually cooled down by the parting with its caloric; or, perhaps, when preserved from ebullition, though at an intense temperature, by the weight of the solid noncrystalline overlying strata (schists, sandstones and limestones), which it supports, and raises by its own elevation); in such a case, the rock will cool down superficially, gradually transmitting its caloric to the outer strata; and its contained vapours, if any, be by degrees wholly condensed. It will be then entirely compact, and void of pores, and preserve nearly the same characters as previous to its elevation.

It is, I think, probable, that many very considerable masses of crystalline rocks occurring on the surface of the globe, for instance, many granites, sienites, and perhaps porphyries, and even some trachytes, have been elevated to their present situation in this manner. The facts confirmatory of this idea will be found in a subsequent chapter of this essay.

§. 10. The vapour that escapes by percolation from lava masses through the pores and crevices of the outer and already consolidated part, is generally found to be accompanied by a variety of mineral substances, in a state either of solution, or sublimation; or finally as permanent gasses. These substances, it has been already observed, are not discoverable in the vapours that rise in immense volumes from lava immediately upon its protrusion; and which appear to be either purely aqueous, or to contain but a very small proportion of mineral matter. But, as the quantity of steam evolved diminishes, and is at length reduced to thin columns, or *fumarole,* that are exhaled from the narrow and intricate crevices of the superficial lava, the proportion of extraneous substances that accompany it is found to increase. Upon coming into contact with the external air, a certain proportion of the steam is condensed by refrigeration, and deposits most of the substances it holds in solution, on the sides and edges of the fissure. The principal of these are the sulphuric, muriatic, and carbonic acids, either in a state of purity, or combined with various alkaline or earthy bases.

Amongst the latter, the sulphates of lime and ammonia, soda and potass, the muriates of soda and ammonia and the carbonate of soda predominate. It appears probable that the first of these acids is occasionally, if not always, formed on the spot, by the union of a part of the sulphur, conveyed in the aqueous vapour, with the oxygen of the atmosphere. The remaining sulphur, when it is produced abundantly, is deposited in a state of purity, in angular or octohedral crystals, or, when most copious, in stalactitic concretions, at the edges of the orifices, and those neighbouring points where the vapour in which it is suspended, undergoes condensation.

The compounds of boracic acid are rare.

The metallic sublimations are condensed in the same manner on the sides of the fissures of escape. The most frequent of these is specular iron. The muriates of copper and iron and sulphuretted arsenic are occasionally found in similar situations.

It is no easy matter to determine the origin of these different substances, and to distinguish those which were originally contained in the lava, and are merely volatilized from the expansion afforded them by the formation of the fissures of retreat, and the escape of the elastic vapour by which they were repressed, from those which are the product of the different chemical combinations operated during the consolidation of the lava, between its various elements and those of the water, which may be partially decomposed during its percolation. Soda, potass, lime, and iron, are constant ingredients in almost every variety of lava. Chemical analysis has occasionally discovered in them, ammonia, bitumen, and muriatic acid. Sulphur in combination with iron or copper is also frequently present.

These substances when brought into contact by the motion and aggregation of the lava at an intense temperature, will necessarily act upon each other, and give rise to various combinations which rise in a vaporized state, or are dissolved instantly by the aqueous vapour, which ascends with them from the interior of the lava.

The intense heat of the lava in the interior of the mass is sufficient of itself to effect the decomposition of some of its ingredients, as the pressure occasioned by the expansive force of the interstitial fluid is diminished by its expansion and partial escape through the pores or fissures of the solid surfaces; and the progressive increase of these mineral compounds in the exhaled vapours, as the lava cools down, that is, as the elasticity of its confined vapour is diminished, warrants the conclusion that such a decomposition is, in reality, the immediate cause of their production. The specular iron, so frequently met with in lava-rocks, is evidently sublimed by this action, and it is remarkable that it is always found in the upper parts of the bed or current; and, moreover, that where the fissures or interstices of any lava rock are lined with an abundance of this mineral, the lower parts of the rock do not influence the magnet, having obviously lost all, or the greater part of their iron by sublimation. In many of the currents of Ætna, the upper parts, which are at the same time very porous and cellular, contain much specular or oligistic iron; while on the contrary the lower and compacter division abounds in magnetic iron, in grains, or octohedral crystals. Here the magnetic iron originally contained in the central and upper parts of the current, has been evidently volatilized, and deposited again in the form of specular iron, while that of the lower part, from which little or no vapour was enabled to escape by percolation, remains unchanged.

In the same manner the other mineral ingredients of a lava are occasionally volatilized, and again deposited, not unfrequently under new combinations and forms in the cavities of the lava, as its temperature is gradually lowered.

This was in all probability the origin of those delicate, and often capillary crystals of hornblende, augite, melilite, and other minerals, which occur in the cellular cavities of many lava rocks. (Capo di Bove, &c.)

Crystals of augite are also said to have been formed by sublimation in the interior of some of the houses of Torre del Greco, destroyed by a current of lava in 1794.*

The substances taken into solution by the aqueous vapour are often also deposited in a similar manner, either in crystals, or stalactitic and mammillated concretions, which line the interior of the vesicles, or the interstices of the mass. The minerals found in this situation are principally either siliceous, or calcareous, consisting of carbonate of lime, calcedony, fiorite, and some of the numerous varieties of the zeolite family. They all bear the character of having been deposited from aqueous solution. We know indeed that water at a high temperature, and particularly with the assistance of soda, readily takes both silex, and carbonate of lime, into solution.

---

* Brieslak Voyage dans la Campanie, tom. i.

It is obvious, therefore, how the aqueous vapour, in its passage through the lava, whether by percolation, or in the form of bubbles, may become impregnated with either, or both, of these ingredients; and when prevented from any further escape, may force these substances, as it is gradually condensed, to crystallize on the sides of the containing cavity.

Sometimes the water remains in the centre of this geode accompanied by a portion of some permanent gaseous fluid, as in the Calcedonies of the Monti Berici, and in the cavities of numerous quartz crystals in granitic rocks. In general the water has escaped, probably by filtration, towards the lower parts of the mass after these had been entirely consolidated. During this filtration, it perhaps continued to deposit the substances it still retained in solution in the cavities, pores, and minute fissures, through which it slowly percolated; and hence, it probably is, that many of the cellular cavities of these rocks, have been *entirely filled* with the calcareous or siliceous deposit; which could not have been all held in solution by the minute quantity of steam, to whose expansive force the original formation of the vesicle was owing.

In many cases the subsequent filtration of water, carrying in solution mineral particles from *overlying* rocks, may have produced the same effect. This is particularly credible with respect to the amygdaloidal basalts, which are found so frequently underlying calcareous strata, or in such positions as make it probable that they were once covered by them, and which therefore were consolidated beneath an ocean of water holding calcareous particles in suspension.

The sulphuric and muriatic acids which are evolved in a state of freedom, when deposited by condensation on the lava which forms the edges of the crevices of escape, act upon its substance, and give rise to fresh changes. The sulphuric acid uniting with the alumine of these rocks, produces a sulphate of alumine, which is often carried away by the rains and deposited in vast accumulations in lower situations. The mining operations of the different alum-works of Hungary, Italy, &c. are carried on in deposits of this nature.

The sulphuric acid uniting with the lime, produces a gypseous efflorescence, which is often found coating these decomposed lava-rocks, in considerable quantities. The iron of the lava is either attacked by the acid, and aggregated into crystalline, or concretionary pyrites, disseminated through the disintegrated rock; or assuming various degrees of oxydation, stains it with stripes of different shades of brown, yellow, red, green and blue. The silex alone remains untouched, and when all, or a large proportion, of the other ingredients have been washed away, the lava appears occasionally changed into a light, harsh, carious, and highly siliceous rock, bearing no resemblance whatever to its original character. Where the exhalation of vapours strongly impregnated with this acid have continued for a great length of time, the change thus effected in the neighbouring rocks are, from the continued shifting of the fumaroles, extensive and remarkable, and have acquired for such spots the common appellation of Solfataras or Souffrières. These usually are found in the interior of a volcanic crater, as might be expected; since the evolution of vapours from a current of lava that has flowed away from the volcanic vent upon the surface of the earth, must be extremely limited in quantity and duration; whereas, after most eruptions, a vast body of heated lava must remain beneath the bottom of the crater, in which the process of solidification goes on slowly for an indefinite, and often very long period, affording a continual source of aqueous vapour, charged more or less with various mineral substances.

§. 11. When the superficial parts of the mass of lava undergoing consolidation have cooled down to the temperature of the atmosphere, or when its vapours escape through fissures in solid overlying rocks of any heterogeneous nature, and at a low temperature, they are more or less condensed by refrigeration, before they issue into the air, and make their appearance in the shape of springs of water at a variety of elevated temperatures, occasionally even many degrees above the boiling point.

These hot springs are common in all volcanic districts, and in those where any more energetic external development of volcanic action has been long wanting, are the only indications of the still remaining activity of the focus below. There can be no doubt that the quantity of caloric which is enabled to pass off through permanent fissures, in this way, materially contributes to maintain the outward tranquillity of this subterranean focus.

It may be even conceived highly credible that this regular and placid transmission of caloric, may be exactly proportioned to the supply constantly communicated from below to the volcanic lava-reservoir; and by thus preserving the focus at an uniform temperature, may entirely prevent that accumulation of heat, by which alone fresh eruptions could be produced.

The quiescence, or what is usually called the extinction of these volcanic foci, will then have been occasioned by the production of the fissures through which the excess of caloric is thus enabled to escape in combination with water.

The water of thermal springs holds in solution the various mineral substances, which, in the form of vapour, it carried upwards from the heated lava below.* Some of these are deposited immediately on its coming into contact with the air, while others remain permanently combined with it. The siliceous sinter deposited by the hot springs of Iceland is well known, as well as the calcareous incrusting springs of numerous volcanic districts, (Auvergne, La Tolfa, that of the bridge of the Incas in the Andes, &c.)

Even when the latter sources no longer retain any elevation of temperature, and derive from rocks which do not present any indications of recent volcanic action, there is still reason for supposing them to have the same origin as the hot springs, but to have parted with more of their heat by a longer passage through cold rocks, and perhaps by their mixture with other superficial springs of atmospheric origin. (Derbyshire, &c.)

Where on the contrary the traces of volcanic action are recent, and the intensely heated lava very close to the surface whence the springs issue, the curious phenomena of intermittent fountains are sometimes produced, such as afford so magnificent a spectacle in the Geysers of Iceland.†

Hot springs of a similar character are also produced by the filtration of atmospheric water through the crevices or interstices of a mass of lava not yet cooled. An example of this occurrence on the large scale is afforded near the volcano of Jorullo in Mexico. The rivers Cuitemba and San Pedro which lose themselves on one side, beneath the vast sea of

* From the heights of the volcanic mountain of Puracè (Peru,) there flows a rivulet so strongly impregnated with sulphuric acid that the inhabitants call it Vinegar river, and no fish are found in it till four leagues after its junction with the Cauca.

The other volcanos of Popayan and Pasto evolve also copiously hot vapours abounding with sulphuric acid, and accompanied by sulphuretted hydrogen gas. (*Humboldt, Voyage Historique.*)

† The theory of these springs has not been very satisfactorily explained. The following considerations will perhaps sufficiently account

lava that forms the Malpays,* issue on the opposite side, as permanent springs of large bodies of water, at an elevated

---

for their phenomena. The lavas of Iceland are replete with hollow blisters or caverns. Let us suppose C such a cavity in a vast bed of heated lava, which may be either isolated and cooling slowly, or connected with the volcanic focus below, and receiving a constant supply of caloric from

Fig. 18.

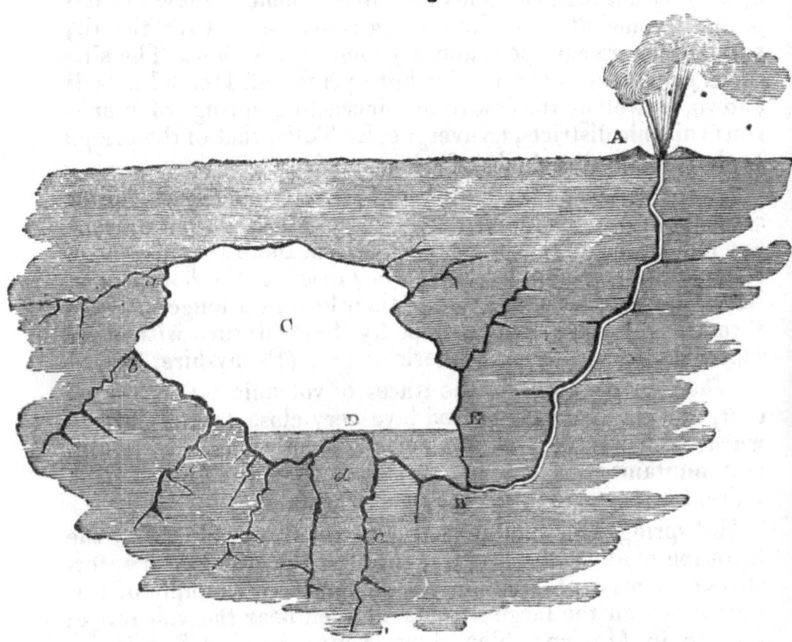

thence. The steam emanating from numerous fissures, *a*, *b*, *c*, *d*, *e*, &c. collects in C and is partly condensed into water by the pressure of the column in the fissure of discharge, A B. The increasing temperature of the floor and sides of the cavity, and the accumulation of steam evolved by the feeders, *a*, *b*, *c*, *d*, &c. augment the expansive force of the steam collected in C till it at length overcomes the resistance of this column of water A B, and discharges it from the orifice A. As the water issues from A, the pressure on the steam in C is diminished, more of the water D E is vaporised, and the ratio of its expansive force to that of repression in the column A B, increased. The jets therefore that are thrown up from A must augment in violence. As the water decreases still more in quantity, some of the steam

---

* Vide Appendix, No. 2.

temperature. When Humboldt visited Jorullo in 1804, forty-six years after the eruption by which the lava was produced, their temperature was 52 cent. but I understand, from the authority of Mr. Bullock, jun. who has recently visited the spot, that they have since that time been lowered considerably, in consequence of the refrigeration of the lava bed through which they flow, and are now scarcely more than a few degrees above the mean temperature of the air.

At Bertrich-bad, in the territory of Luxembourg, a thermal spring rises immediately below a spot on which three proximate vents of volcanic eruption were formed at no very distant period; and probably owes its qualities to having percolated through part of the focal lava reservoir of these eruptions, not yet entirely refrigerated. It may be suspected that the temperature and mineral nature of this spring has progressively diminished. It is at present below blood-heat, and has as nearly as possible the taste of pure fountain water. In earlier times its thermal and mineral qualities must have been more striking to have acquired for it the great reputation which it formerly enjoyed as a bathing-place.

§. 12. The permanent gasses that emanate from the crevices of lava, during its consolidation, are less palpable, and not so easily recognized, mixing of course immediately with the atmospheric air, and leaving no trace of their existence.

Those which have been occasionally observed are principally the carbonic acid, azote, and sulphuretted hydrogen gasses.

The first of these gasses was detected by MM. Monticelli and Covelli, in the exhalations from the fumarole of the lava of Vesuvius while still visibly incandescent. The same gas, mingled perhaps with azote, frequently escapes in very considerable quantity from numerous crevices in the sides of a volcanic mountain immediately after the termination of an erup-

---

escapes together with it, and at length, when all the water has been driven out, the whole remaining body of steam issues in one powerful burst.

But by the evaporization of all the water it contained the temperature of the cavity C. is reduced to the boiling point in the pressure of the atmosphere alone; and its sides proportionately refrigerated. The vapour therefore, that rises from the feeders is condensed into water, and collects at the bottom of the cavity in B. This water rises in the pipe A B, as the steam accumulates at C, preserving an equilibrium with its expansive force, until the equilibrium is broken by this latter having acquired sufficient power to discharge the water from the orifice of the crevice A, when the phenomena of aqueous eruption recommence.

tion, that is, when the crater or central aperture of escape is completely closed.

These mephitic emanations are extremely destructive to vegetation, and do great injury to the plants growing near the points from which they proceed. During the great eruption which convulsed the island of Lancerote in 1730—1734, these exhalations appear to have been equally deleterious to animal life; all the cattle of the island are represented to have been suffocated by them. M. Hubert, as quoted by Bory de St. Vincent, mentions that during an eruption of the volcano of Bourbon, he observed seven or eight birds, flying at gun-shot height above the current, drop suddenly, as if "asphixiès," upon its surface the moment they entered the cloud of vapours that emanated from it. This circumstance lends some colour to the fables related of lake Avernus, which is certainly a volcanic crater of no very ancient date, and from whence carbonic acid and azotic gasses may once have been so copiously exhaled (rising through the atmosphere in consequence of their elevated temperature) as really to affect the birds that flew over it. It is much more probable that this fable should be founded on a fact of this kind, than that any fortuitous invention of a poetic imagination should coincide so well with known phenomena. The banks of the Lago D'Agnano in the immediate vicinity of the Avernus still exhale carbonic acid gas, as well as sulphureous vapours, on many points.

§. 13. The period that intervenes between the deposition of a bed of lava upon the surface of the earth, and its complete refrigeration, that is, its attaining the mean temperature of the atmosphere, must be determined by the combinations of numerous circumstances, such as

I. The figure of the mass. It is obvious that the more equal its dimensions in every direction, the longer will the central part retain its heat; and, vice versâ, the greater the superficial extent of the bed in proportion to its thickness, the more rapid will be the process of refrigeration,—other circumstances remaining the same.

II. The situation of the bed; by which it is more or less exposed to the influence of refrigerating media—such as currents of water or air, &c.; and the conducting powers of the solid masses with which it is in contact.

III. The tendency of the lava to part more or less rapidly with its caloric, either in combination with aqueous vapour or by radiation—dependent upon its compactness, and perhaps its mineral composition.

We have already seen that the coarser the grain of the lava and the more irregular its disposition, the greater the quantity of vapour that is enable to escape by percolation; and the more rapid therefore, ceteris paribus, will be its solidification.

The loss of caloric by radiation follows perhaps the inverse ratio, since the conducting powers of the mass will be probably proportioned to its compactness, and therefore to the fineness of its grain, or the regularity of its disposition.

Under favourable circumstances a body of lava will retain an intense heat and liquidity in its interior, during a very long period.

Sir W. Hamilton lighted small strips of wood, by inserting them into the fissures of a lava current from Vesuvius, nearly four years after its emission. Currents of Ætna are mentioned by Ferrara and Dolomieu as still moving down the sides of the mountain ten years after the eruption by which they were produced, and other as emitting vapours 26 years after their escape from the volcano. In the case of Jorullo, already cited above, a massive bed of lava appears to have retained an extreme internal heat, attested by the exhalation of steam in considerable quantity from numerous fissures in almost every part of its surface, till within a very few years, though the eruption by which it was emitted dates from 1759,

CHAP. VI.

Divisionary Structure assumed by Lava on its Consolidation.

---

§. I. Both the partial escape and gradual refrigeration of the interstitial fluid, by diminishing the elastic spring of what remains in the lava, brings its solid particles into closer union.

As they approach within the due distances, their mutual attractive forces are developed, and cause them to aggregate together (condensing the intermediate vapour) into a solid mass, which will be more or less porous according to the size of the crystalline particles, and the previous irregularity of their polar arrangement. This aggregation cannot take place without occasioning a shrinking, or diminution of volume.

If this process commence in the centre of a liquid body, which will yield from all sides towards the part that contracts, no separation of parts will ensue. Such is the case when a mass of lava is consolidated by external pressure; the globular concretions then formed are not separated by any crevices.

But if the process of consolidation begin at the surface of a mass, as is always the case when it is effected by contact with either a rarer or colder body, the tendency to contract is exerted with equal energy on all points of the then superficial layer which is undergoing the process. The action of the contractive force in a direction perpendicular to the surface, effects no rent; since the lava that remains much more perfectly liquid beneath this surface yields or subsides freely in that direction; but on the contrary, that portion of the contractile force which is exerted in the plane of the surface, or in one parallel to it, is opposed on any one point by the contemporaneous shrinking of all the surrounding parts.

By the action of these opposite forces the layer must be divided, by a greater or less number of rents, into distinct portions, in each of which the individual force of contraction overcomes the opposite contractile forces of the proximate parts. In each of these portions, therefore, a centre of attraction establishes itself; the crystalline particles that occupy the centre remaining stationary, while those that surround it are drawn more or less from all sides towards the centre. The rents or fissures of retreat will be obviously perpendicular to the plane of the surface at which the consolidation commences, and are produced along those lines in which the contractile forces of the proximate centres of attraction balance one another. If the process was perfectly contemporaneous throughout the whole layer, and its substance completely uniform, it is clear that the points on which the centres of attraction establish themselves would be symmetrical and equi-distant, and their contractile forces perfectly equal.

In this case all the spheres of attraction would be equally similar in size and form, and would arrange themselves as closely as possible, that is, in the manner of the cells in a honey-comb, or as the circles in fig. 19.

Fig. 19.

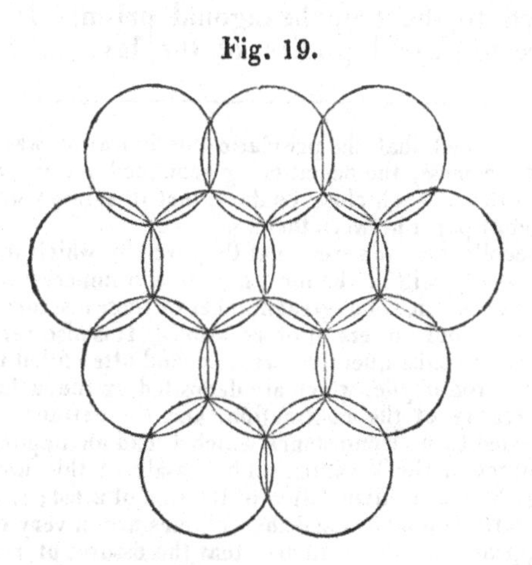

The fissures of retreat produced by the contractile force of all the spheres, acting contemporaneously, must evidently therefore divide the consolidated layer into hexagons; each straight fissure being tangential to the opposite spheres of attraction between which it is formed.

§. 2. But since the liquidity of the lava beneath this superficial layer allows it to yield to the contractile force acting in a direction perpendicular to the surface, no fissure will be produced parallel to that surface; but by the continued propagation inwardly of the process of consolidation, the fissures of retreat already formed will be prolonged towards the interior, and must divide the mass into so many hexagonal prisms; in each of which the line described by the centres of attraction forms the axis.* The diameter of the hexagons, or the distances of the proximate centres of attraction, will vary inversely with the intensity of the attractive force, and therefore, with the circumstances which favour the activity of this force, viz. the slowness of the consolidating process, and the mobility of the solid particles. But in all lavas the roughness and irregularity of the outer surface prevent the uniform propagation of the process of consolidation in true planes; and their substance can never be completely uniform; hence constant irregularities must arise in the diameter, and in the number and length of the sides of the polygonal prisms into which the mass divides itself; it is, however, a well known fact, that their figures usually oscillate about the hexagon; and that the more well-defined the columns, the more nearly do they approach to the true hexagonal prism. It is obvious that the greater the liquidity of the lava, and the more

* It has been denied that the prismatic configuration was produced by contraction of the mass, the denial being grounded on the assertion that the prisms are often in such close contact, that it is impossible to insert a knife or a sheet of paper between them.

It must be recollected, however, that the force by which the contraction is produced acts only within the most minute and imperceptible distances, and that a space consisting of a great number of these distances, may yet be almost insensible to our powers of observation. It is also certain that the fissures of retreat are subsequently narrowed, and often filled up by infiltrations, usually of iron oxyde, which are deposited on the surfaces of the columns. The reality of the contraction, and the extraordinary force it exerts, are attested by a circumstance which I had an opportunity of observing at Burzet in the Vivarais. The basalt of this locality contains numerous large knots of olivin, often of the size of a fist; it is at the same time very regularly columnar, and the columns are in very close contact. Yet it has happened in many instances that the fissures of retreat have divided one of the large olivin nodules into two; half being enclosed in one column, and the other half in another proximate column. Though the division is complete, and the fracture surfaces smoothed probably by time, and aqueous filtration, they correspond so completely that it is impossible to doubt their former union in the same nodule. Their separation can only be accounted for by a powerful contractile force exerted during the formation of the columns.

homogeneous its substance, the more uniformly will the force of attraction be dispersed through it; and also that the more slowly the consolidation is propagated, the greater will be the regularity with which the centres of attraction arrange themselves. And this accords entirely with the known fact, that the most perfect columnar division has always taken place in the *interior*, and generally in the *lowest* part of the lava current, and particularly where any considerable accumulation of the liquid has been formed in some concavity of the underlying surface; that is to say, exactly on those parts which cooled most *slowly*; and again, that the regularity of the columnar structure is, ceteris paribus, proportioned to the fineness of grain, and homogeneity of the lava; since, wherever the consolidation was suddenly effected, the prismatic masses circumscribed by the fissures of retreat must be rude, shapeless, and iregular. This will uniformly be the case with the outer and exposed parts of the current; and when the grain of the lava is extremely coarse and irregular, it will happen even in the interior, since its solidification will be rapidly effected by the percolation of the elastic vehicle through the pores and fissures of the upper part.

Hence has arisen the error which has so long and so generally prevailed of supposing the prismatic configuration to be solely confined to the earliest lava-rocks. In reality this structure is common to the *interior* of the lava masses of every age whose composition and circumstances of liquidity and disposition were favourable to its production. Such masses, when of an early date, have been worn through, denuded, and excavated, by long exposure to aqueous erosion, and other destructive processes; owing to which the prismatic configuration of their interior has been disclosed to view; while, in the currents of modern volcanos, such opportunities of observing their internal configuration are more rarely afforded. But where this *is* the case, as in the sections afforded by the deep ravines, or vertical cliffs of Ætna, Iceland, Bourbon, and even Vesuvius, the most regular columnar structure is often observable, even in lavas of which the date is known.*

The consolidation of lava, occupying fissures of a solid rock, (dykes or veins,) must take place entirely by the condensation of the interstitial fluid, and will be much more

* The basaltic rock of La Motte, near Catania, is regularly columnar in the interior, and exhibits this structure on one side, where it has suffered great degration from the waves of the sea; the other sides, as well as the surface, are amorphous. Ferrara, Campi Phlegrei, p. 322.

slowly effected than that of a body of lava exposed superficially to the contact of air or water. Hence its prismatic configuration will be far more regular in proportion to the size of its grain. The prisms in this case will, of course, be perpendicular to the sides of the fissure.

The correctness of this reasoning is proved by the regularly columnar division of numerous dykes in basaltic countries.

As the process of solidification propagates itself towards the interior of the lava-mass, the fissures of retreat, and consequently the columnar prisms they circumscribe, are prolonged in a direction perpendicular to the outer surface, or at least to the plane in which the process of solidification acts at any one moment of its progress. If any circumstances change the direction of this plane, or produce irregularities in the propagation of the process, the direction of the columns suffers a proportionate deviation.

The columns will be usually more or less perpendicular to the surface on which the solidifying process first acts; (as is the case in the desiccation of starch, clay, &c.) Therefore, when they are formed upon a convex surface (as *a c*) they will

Fig. 20.

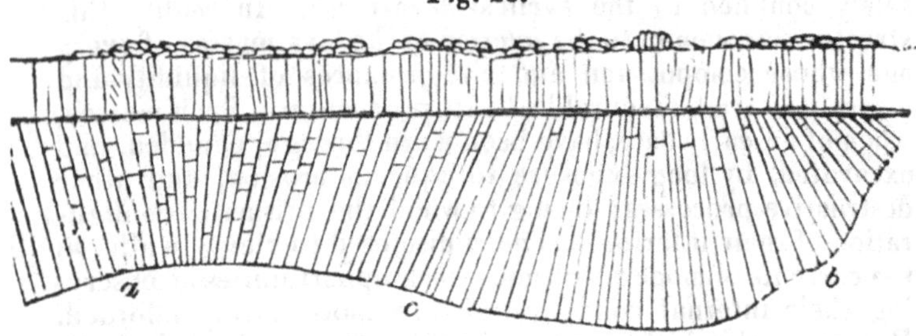

diverge towards the exterior of the mass; when upon a concave surface, (as *c b*,) they will converge in the same direction.*

* One of the main causes of the rapid degradation of volcanic rocks, particularly basalt, lies in the facility with which the rain-water penetrates between its prismatic divisions, and the consequent effect of frost in rending them asunder. It is obvious that this disaggregation will be greatly facilitated by the prisms converging downwards, owing to which their own weight assists in disuniting them; while it is on the contrary most effectively opposed by the contrary arrangement.

§. 3. But the prolongation of the columns is dependent on the lava of the interior always yielding to the contractile force, which acts in the direction of the columns themselves, or allowing of the subsidence of the outer solid crust in that direction.

If its liquidity should be too imperfect for this, or the supply of liquid from a higher source wanting, the columns are themselves divided by transverse fissures, or *joints*, which separate them into so many portions or articulations.

The frequency of these joints will vary with the difficulties opposed to the supply of liquid lava in the interior, or to the subsidence of the part already consolidated; and, therefore, the length of the columns without a joint is favored by an excess of liquidity, and consequently by fineness of grain. In the same manner the columns that form within any accumula-

---

This consideration explains the origin of those insulated conical peaks of frequent occurrence in basaltic districts, and which have been so often appropriated to the sites of feudal fortresses; these will for the most part, if not universally, be found to owe their existence to the extreme stability occasioned by a pyramidal grouping of columns, and the freedom from joints, which has enabled them to outlive the destruction of the remainder of the beds to which they belonged.

One of the most magnificent columnar peaks of the kind is to be seen at Murat (Dept. Cantal) in France. See pl. II. fig. 4.

A segment of the basaltic current, of which this group once formed a part, is still visible in the distance.

The general nature of this fact suggests the important remark, that the destruction of the basaltic beds in which such conical eminences were once included, has been the result of the slow operation of the causes now in action. Any violent rush of waters, of the nature of a deluge, would sweep away a pyramidal cluster of columns as easily as a vertical one; it is the long continued alternation of rain and frost alone that will explain the uniform destruction of the one and preservation of the other.

If the insulation of such peaks has been produced by these apparently powerless agents, it will follow that they are capable of effects of greater magnitude than is yet generally allowed, and their influence in the destruction of other rocks has been probably far greater than many Geologists of the present day are willing to concede.

Wherever seams of stratification or divisionary structure exist, water will introduce itself by capillary attraction, and the expansive power exerted by its crystallization is known to be immense. When a mass has been in this way loosened, it is easily displaced; either by its own gravity, or an external impulse such as that of water.

Another frequent cause of dilapidation, particularly amongst volcanic rocks, but which must not be enlarged on in this place, is their super-position to beds of clay, tufa, scoriæ, and other loose matters, which are easily washed away, and the overlying rock consequently undermined. The prismatic structure, by allowing the percolation of water to these lower beds, obviously accelerates the process.

tion of lava on a concave surface, should be long in proportion to the supply of liquid lava from the higher parts of the current, and those formed on a convex surface short for the opposite reason (see fig. 20).

By an extreme frequency of joints, the columnar tends to pass into the globiform structure.

§. 4. The distance of the proximate centres of attraction, it has been said, varies inversely with the circumstances that favour the energy of the attractive force; and this accords with the fact, that the finest-grained lava-rocks usually present the smallest concretions.

When the circumstances of texture and composition are extremely favourable to the increase of the attractive force, more than one concretionary separation of parts may take place successively in the same mass.

The first may accompany the diminution of bulk which ensues from the escape of a part of the elastic fluid by percolation.

A second series of fissures may subsequently be formed, perpendicular, or nearly so, to the first, when the temperature of the interstitial fluid, which cannot escape, is so far lowered by slow loss of caloric, that its expansive force yields to the mutual attraction of the neighbouring particles. In this manner the prisms first formed may be divided into numerous smaller prisms perpendicular to their bounding surfaces.

The former will be large and very irregular, owing to the rapidity of the formation. The latter small and more perfect.*

§. 5. It may occasionally happen that the lava refuses to yield in any direction towards the centres of attraction; in this case the fissures of retreat will be equidistant on all sides from their centres, and will circumscribe a polygonal mass approaching in form to an angular spheroid, (globiform structure.) Subsequent contraction may divide this mass either into prisms having their axes perpendicular to the surface of the spheroid, and therefore converging to its centre, or into concentric lamellæ; according as the contractile force, in its progress from the superficies to the centre of the mass, expe-

* Amongst the basaltic colonades of the Vivarais are some beautiful examples of this mode,—A part of one may be seen in fig. 21. It is remarkable that the lowest part of the bed in which the columns are regular, large,

riences least resistance in the direction of the radii, or of the tangents, to the sphere.

The former will require a high liquidity, and therefore fineness of grain in the lava; the latter will take place where the mobility of the particles is very imperfect.

At St. Sandoux, in Auvergne, is a beautiful example of the former variety of divisionary structure, in an enormous spheroid of basalt, more than 50 feet in diameter, composed of very regular and jointed columns diverging from its centre, where they are closely united, to the circumference, where a considerable space is left between them.*

The division into concentric lamellar spheroids is common and well known.

§. 6. During the advance of the concretionary process through those parts of a lava-bed in which the exercise of the

---

long, and almost always vertical, is separated by a decided line from the upper and less regularly columnar part, in which there is generally a double

Fig. 21.

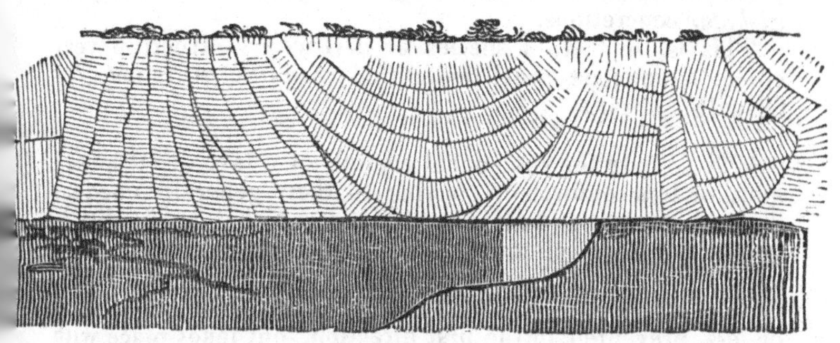

system of concretions—this line runs always very evenly, and parallel to the lowest surface of the current, without regard to the irregularities of thickness in the upper part. This inferior division, I am inclined to think, parted with its caloric downwards to the substratum, and the line in question appears to have been the separation between the parts which lost their heat by transmission upwards, from this. The lower cooled solely by *condensation* of vapour, the upper both by *condensation* and *escape by exudation*. Hence their difference of regularity.

The prismatic trachyte of Ponza and Palmarola offers examples of a similar nature.

*. See an engraving of this rock in Faujas de St Fond, Sur le Vivarais et le Velay.

attractive forces is most favoured by the extremely slow refrigeration of the interstitial fluid, it may happen that some of the proximate crystalline particles of the prevailing mineral will be enabled to *overcome the resistance that opposes any change in the direction* of their poles, and unite into large and regular crystals. Such an arrangement appears to have produced many of the long and delicate crystals of felspar or hornblende, which are found imbedded in some of the compactest and finest-grained basalts, and which appear to have crystallized since the settling of the lava, and thus to form an exception to the otherwise general remark of the pre-existence of the component crystals of such rocks.

When the lava is extremely comminuted, in the place of crystals, globular concretions, or crystallites, are produced by the same partial reversion of the poles in neighbouring particles.

This arrangement must occasion a diminution of volume, which if not compensated by subsidence, or a supply of liquid lava from without, will produce numerous separations of continuity, which taking effect in those planes in which the attractive forces of the neighbouring centres balance one another, divide the mass into numerous *polyhedral* or *angulo-globular* concretions.

This is, in fact, a species of divisionary structure very common amongst pearlstones, and variolitic greystones, or felspathic basalts.

§. 7. Where the component crystals of a lava are disposed more or less conformably, owing to their generally unequal dimensions, it has been already observed that its fluidity is great in the direction of their longest plane surfaces, and very little, or entirely null, in the transverse direction. Owing to this cause the formation of retreat fissures is, more or less, prevented in the first direction, but takes place with the greatest frequency in the latter; and thus the lava is divided into concretionary plates or lamellæ, of greater or less thickness, according to the greater or less difference in the fluidity of the lava in these opposite directions. When the thickness is comparatively great, the structure is called *tabular;* when less so, *lamellar;* and when the plates are very thin, *slaty* or *schistose.*

These different structures are, as might have been expected, abundantly exemplified in varieties of the clinkstone or scaly lava-rocks, both felspathic and augitic; and the same cause may not improbably be looked to, to account for the flat prismatic divisionary structure of syenites, granites, gneiss, and the earliest traps.*

This structure is often accompanied by the columnar, of which it is wholly independent. The latter being produced *first* by the earliest diminution of bulk, while the lava possesses a certain degree of fluidity in all directions. The ruptures occasioned by this are as usual perpendicular to the exposed surface, and therefore generally (though not necessarily) transverse to the planes of the lamellæ; which being always parallel to the direction in which the lava moved, are also, of course, mostly parallel to the outer surfaces of the bed, current, or dyke.

A remarkable example of this united structure occurs in the Roche Tuiliere in the Mont Dor. The clinkstone of which this rock is composed (an insulated fragment of the vast current descending from the Puy Gros), is regularly divided into nearly vertical columns. It is also extremely schistose, as the vulgar name of the rock implies, the laminæ being used as slates for roofing.

Fig. 22.

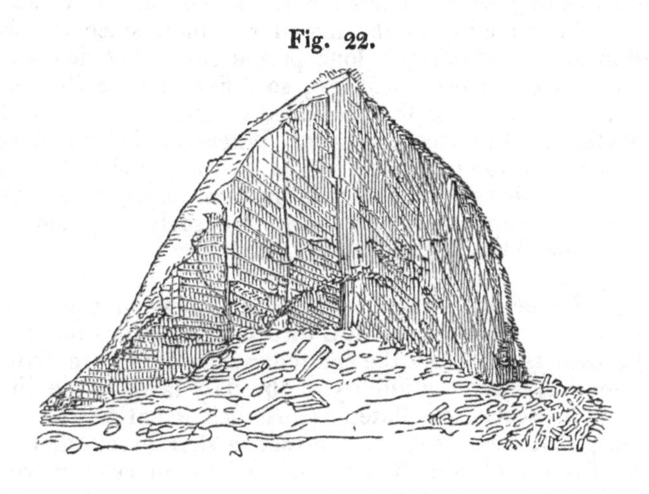

* In fact, the *stratification* of rocks of all kinds, where the strata are separated by *seams*, is produced by this concretionary process.

The direction of the lamellæ in the centre of the rock is horizontal, and therefore perpendicular to the axis of the prisms, but gradually declines towards the North, until it becomes parallel with their axis; the fact, that the former arrangement communicates great solidity to the prisms, while the latter affords an easy access to the agents of division and degradation, perfectly accounts for the isolation of this rock; the part, in which the latter arrangement prevailed, being wasted away.

This curvature of the lamellæ appears to be owing to some involution of the current while flowing, owing perhaps to the intervention of some obstacle, on its fall down a steep slope.

During this lamellar concretionary process a great many of the proximate crystalline particles of the prevailing minerals must be expected to reunite into large and complete crystals; since, in this case, the particles are already more or less conformably arranged, and require therefore little or no change in the direction of their poles to recrystallize completely. In this manner were probably formed the great proportion of the imbedded crystals of clinkstones and slaty lava-rocks in general. Their longest dimensions are necessarily in the direction of the lamellar planes, since the free motion of the crystals is alone permitted in that direction; and their excessive *compactness*, so different from the loose and fibrous aspect of the imbedded crystals of most earthy trachytes, is thus accounted for. Where the heterogeneous minerals were too intimately mixed, or the liquidity too low, to allow of this complete recrystallization, imperfect globular concretions were produced; as in the scaly greystones of Auvergne and the Mittel Gebirge, &c.

§. 8. In the same manner as the incomplete supply or subsidence of the lava in a vertical direction, occasions the transverse fissures by which the columnar concretions are divided at intervals, when the fluidity of the slaty lavas in the direction of the scales, or plates, is insufficient entirely to compensate all the tendency to contract in that direction, transverse fissures of retreat are produced at intervals more or less distant; and by the frequency of these, the tables assume a cubical or rhomboidal form, such as is conspicuous in the earlier traps, in syenite, and even in granite.

§. 9. It has been seen that the Divisionary Structure, or regularity of internal form, which a mass of lava may assume during its refrigeration, according to circumstances comprehends the following classes:

1. The Prismatic or Columnar.
2. The Tabular, Lamellar, and Schistose.
3. The Rhomboidal or Cubiform.
4. The Globiform.
5. The Angulo-globular.

And that two, or even more, of these varieties of form may be taken by the same mass.

§. 10. It is obvious that the Divisionary Structure can only be produced in lavas consolidated by loss of caloric, or of their fluid vehicle. Where they are solidified solely by increase of pressure, their temperature remaining fixed, the concretionary or crystalline arrangement of particles will take place (and still more easily and completely than in the former case, because the process is slower), but no contraction, and therefore no rupture or fissures of retreat, can be produced.

§. 11. As yet I have purposely avoided entering into any discussion of the circumstances which may have given rise to the variety of mineral composition observable in the lavas.

This variety must be supposed owing to one or other of two causes, viz.

1. An original difference in the nature of the subterranean crystalline rock, whose extravasation produced them on the surface of the earth; or,
2. To changes produced in this rock, originally of an uniform composition, during the process of elevation, which was probably accompanied in many instances by repeated intumescence and reconsolidation, before the final emission of the lava.

I own that I incline to the latter alternative, although I am far from being able to explain the mode in which every such change can have taken place.

Our knowledge, indeed, of the laws which determine the formation of different minerals from their elementary ingredients, is at present too limited for the solution of such a question. If, however, we have sufficient evidence of the fact, that the heat and fumes issuing from a mass of incandescent lava, can alter a rock with which it comes in contact under greatest pressure, from a compact carbonate of lime to a highly crystalline dolomite, containing 45·00 of magnesia,* it will not certainly appear incredible that very considerable changes of mineral character may be effected in a mass of lava itself, during the repetition of the processes of intumescence and consolidation, under variations of pressure and temperature; that crystals of augite, or olivine, or leucite, may have been produced from some of the comminuted, and perhaps vaporized, elements of mica and felspar; even that the same original granitic lava may, under some circumstances, have been altered into a trachyte, under others into a basalt.

It would not indeed be difficult to conceive the production of ordinary trachyte, by a very slight change, from a granitic original; for the process of intumescence, when carried far, may easily be supposed to change the felspar crystals from compact to glassy; to dissolve the whole or the greater part of the quartz in the aqueous vehicle; forcing it to assume the crystalline form of felspar on consolidation (as in graphic granite); finally to volatilize the mica, of which part would recrystallize in more perfect crystals, either of mica or augite, on cooling; and a part perhaps give rise to specular or magnetic iron disseminated through the rock, or lining its pores and fissures.

In like manner we may imagine the production of basalt to have been caused by the exposure, within the vent of a volcano, of an intumescent mass of granite to reconsolidation, effected by the augmentation of temperature, and consequent expansion, of its lower beds. In these parts the extreme heat may be supposed to volatilize the mica and other ferruginous minerals, while the intense pressure would separate them in a gaseous state from the felspar, thus leaving a felspathose lava with very little iron in one part of the chimney, and occasioning the crystallization of a highly ferruginous lava in another. The subsequent intumescence and protrusion of these lavas might produce alternate currents of trachyte, clinkstone, or compact felspar, and basalt, or grey-stone.

Where the felspar, or a part of it, was reduced to an extreme degree of minute division, amounting perhaps to com-

* As in the Tyrol (see Von Buch, Lettre a Humboldt); in Sky (see Macculloch's Western Islands, vol. i. p. 325, &c.); in the Isle of Zannone (see Memoir on the Ponza Isles, Geol. Transactions); and in many of the dykes cutting the chalk of the north of Ireland. The well-known experiments of Sir James Hall are confirmatory of these observations.

plete fusion, the circumstances under which the rock was reconsolidated within the focus or vent, may have given rise to the crystallization of its elements in the proportions which characterise olivine or leucite; and other peculiar circumstances, the nature of which it is impossible to anticipate from our incomplete knowledge of the laws of mineral crystallization, and of the effect of high pressures and temperatures in determining it, may have occasioned the casual and local production of the rarer volcanic minerals, such as Haüyne, Melilite, Stilbite, Analcime, Ice-spar, Sommite, Vesuvian, &c.

With regard to the occurrence in Nature of these varieties of lava, the fact is, that they are often, nay generally, found to have been produced successively, sometimes alternately, by the same volcano, or at least from the same system of vents.

Thus in France, in the Mont Dor, Cantal, and Mezen, (three habitual volcanic vents,) currents of basalt, trachyte, and clinkstone, may be seen to alternate.*—See the annexed section of the rocks worn through by the waterfall immediately above Mont Dor les Bains.

Fig. 23.

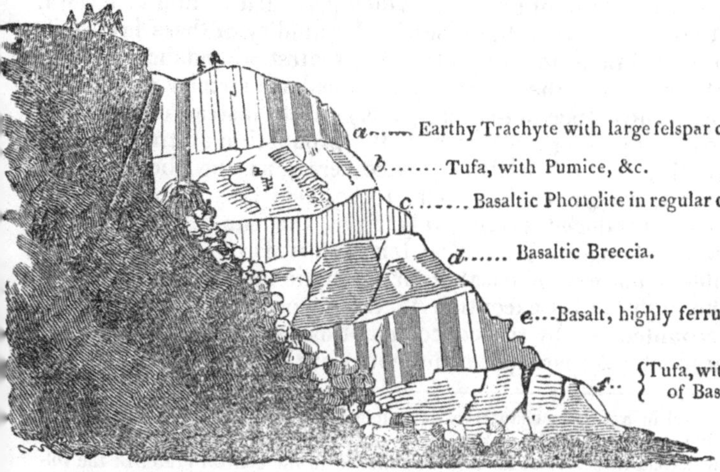

a...... Earthy Trachyte with large felspar crystals.
b....... Tufa, with Pumice, &c.
c. ...... Basaltic Phonolite in regular columns
d...... Basaltic Breccia.
e...Basalt, highly ferruginous.
f.. Tufa, with dykes of Basalt.

Natural Section at the Cascade of Mont Dor, above the Baths.

* It has been justly remarked by Beudant and other writers, that when trachyte and basalt are found within the limits of the same volcanic group, the trachytes form the central eminences, and the basalt is chiefly found at the out-skirts of the group. The superior fluidity of the basaltic lavas,

In the Cantal similar alternations are visible. In the Mezen a very felspathose clinkstone (slaty trachyte) rests upon basalt, and supports in turn a bed of the same rock; and this fact may be observed on many points. In the Chaine des Puys, near Clermont, eruptions of basaltic lavas have almost constantly succeeded those of trachyte, from the same vents. (Grand et petit Puy de Dome.) (Grand et petit Cliersou.) (Grand et petit Sarcouy,) &c.

In the Euganean Hills, in the Monti Cimini, at the Lago di Bracciano, in the Campi Phlegræi, in the Ponza Isles, and Ischia, amongst the Hungarian groups, the Siebengebirge, in Teneriffe, and at Guahilagua and at Xalapa in Mexico, trachyte and basalt occur together, and appear to have been successively erupted from the same or very neighbouring vents.

It is strange that, in the face of these instances, MM. de Humboldt and Beudant should speak of a repulsion or antagonism as existing between the trachytic and basaltic formations.* It is natural, and to be expected, that where basalt has been the latest and most copious production of any volcanic vent, its currents should conceal entirely or partly from our view those of trachyte, which preceded it, and vice versa. Again; the great difference in the fluidity of these lavas will have led them to accumulate in greatest abundance, the one at a distance, the other in the immediate vicinity of the central vent.—(See note, last page.) But, instead of their mutually repelling one another, the production of trachyte preventing that of basalt, and vice versa, as the above writers assert, the *general* law seems to be, that they occur together, being produced successively from the same or proximate vents, though in general at long intervals of time. The complete isolation of basaltic or trachytic formations stands in the light of an exception to the general rule, and may be accounted for by the concealment of so much of the earliest productions (perhaps subaqueous) of the vent.†

---

proved in a former chapter, will evidently account for this general fact. Thus, in the Mont Dor, the trachytic currents have in no instance flowed more than three, or at most four, miles from the central crater of the volcano; the basaltic currents, on the contrary, have reached a distance of fifteen miles or more.

* Humboldt Essai Geognostique, p. 349. Beudant, Hongrie, iii. p.587.

† The same spirit of generalization from pre-conceived hypothesis, has led M. Beudant to assert that all trachytes are of submarine origin. This is evidently false of the trachyte of the Cantal, near Aurillac, which has overflowed a freshwater deposit, and is wholly without foundation in most other instances.

The same able geologists have likewise, perhaps, committed another error in attempting to limit the production of volcanic rocks characterized by varieties of mineral composition, to separate periods of the history of the globe.

According to M. Beudant, all trachytic *formations* are parallel to the secondary strata, and universally prior to the tertiary.

The trachytes of the Cantal and Mezen which rest on freshwater limestone; those of the Montamiata, and Monte Cimini, on the tertiary marls and clays of the Sub-Appenine hills; that of the Euganean hills, which cuts through and covers beds of calcaire grossier, might alone set this question at rest; but it is still more completely decided by the certain fact that the eruptions of many recent volcanos have produced trachytic lavas. Most of the active volcanos of America, particularly Popocateptl, Orizava, Capac-urcu, Cotopaxi, Sotara, and Rucupichinca, are trachytic, and eject pumice. Those of Sumatra, Java, and the Moluccas, also appear to produce principally felspathose lavas. The seas in the vicinity of these islands are represented sometimes, after an eruption, to be covered to a considerable extent with floating pumice.*

The Peak of Teneriffe, which has certainly been eruptive at no very distant period, is trachytic.

Volcano, one of the Lipari isles, when in eruption in 1786, threw up pumice. The Lava del Arso, in Ischia, and that of Olibano, near Pozzuoli, both recent streams, are mineralogically trachytes. The lavas of the volcanos of Bourbon, still in almost permanent eruption, are highly felspathose.

These facts are not unknown to the geologists who advocate the antiquity of trachyte; but they escape their evidence by deciding that recent lavas shall not be called by this name. The term however has been, by its inventor Haüy, and subsequently by Daubuisson, Cordier, &c. restricted to a mineralogical meaning; in addition to which so many difficulties lie in the way of the determination of the age of a volcanic rock, that it would be absurd to make the primary name of any one depend on a character of such uncertainty.

---

* See an Account of the Eruption of Tomboro, in Java.—Journal of the Royal Institution, vol. i. p. 255.

CHAP. VII.

Volcanic Mountains.

---

§. 1. Having thus far investigated the laws which determine the disposition and solidification of lava, when protruded in a liquid form on the surface of the earth, we may continue our examination of the structure of volcanic formations, into which this substance, when consolidated, enters largely.

We have, in a former chapter, discussed the form and compositions of those Cones, or Hillocks, both simple and compound, which are produced by a single eruption of fragmentary matter, from one or a few neighbouring vents; and it has been observed, that the weight and pressure of the lava, ascending within the funnel-shaped cavity of such a cone, often breaks down one of its sides, and allows the lateral escape of the liquid matter, which disposes itself at the foot of the cone, according to the circumstances already enumerated.

If the cone, on the contrary, is sufficiently solid to resist this pressure, the lava will often rise until it is enabled to escape over the lowest part of the ridge of the crater, and pour itself down the outward slope of the cone. In this case a part of it congeals in its descent, and remains fixed as a more or less solid rib or prop to the fragmentary cone.

We now proceed to consider the structure of those larger volcanic hills, which are the produce of repeated eruptions from the same vent. The effect of every subsequent eruption from the same orifice, which has already produced a cone of scoriæ and a lava current, must be to cover these with fresh productions of a similar character; and the repetition of such phenomena will create an alternation of the lithoidal or consolidated lavas, with the conglomerate strata produced by the contemporaneous ejections; these different beds accumulating together, into a mass of more or less magnitude, in proportion to the violence and number of the eruptions by which it was produced.

The original cone thus, by degrees, assumes the size and dignity of a mountain. Those currents of lava, which are able to force their way through the side of this hill, harden into massive buttresses at its foot, or upon its skirts; while those which overflow the edge of the crater, add still more efficiently to the strength of the cone, remaining, as so many solid ribs, interstratified with its looser materials. In this manner the mass becomes gradually more and more fortified; and while the outward pressure of the column of lava, raised by ebullition within the chimney, or central aperture, of the mountain, increases with its growing elevation, the strength and solidity of its sides, and, consequently, the resistance afforded to this force, is augmented in the same ratio.

§. 2. These sides, however, frequently give way to the intensity of this force; but not, as in the case of the similar cones, formed solely of fragmentary matter, where the whole side of the cone is broken down, and carried away at once. Where the cone consists of alternate fragmentary and solid beds, knit together into a compact frame-work, should the force of the lava, ascending in the central vent, and acting like an immense wedge driven from below upwards into the heart of the cone, overcome the cohesion of its sides, one or more fissures will be split through them in a vertical direction; and by these the lava makes its escape, with a velocity and volume, determined by its fluidity, the dimensions of the fissure, and the relative height of the internal column. As the lava flows out of any lateral orifice thus established, the surface of the internal column must gradually fall, until it reaches the level of the orifice of emission.

The stream will then, or shortly after, cease to flow: but will probably be succeeded by jets of the elastic fluid from the orifice by which it is brought into communication with the atmosphere. Meantime, the pressure and intense heat of the lava remaining within the volcanic chimney, perhaps enlarges and lengthens the fissure, so as to afford another opening for the emission of lava at an inferior level. From this point the same phenomena take place, and are perhaps repeated from other openings successively produced, one below another, in the side of the mountain, until the weight of the internal column of lava is so far diminished, that it can no longer overcome the resistance afforded by the solid substructure of the mountain; all discharge of lava in currents then soon ceases; and as the column is still further lowered within the vent, its vapour alone escapes, projecting scoriæ upwards, either from the last formed aperture, or the central crater, perhaps from both; and the eruption gradually terminates in the usual manner.

Observations made on the phenomena of known volcanos confirm this reasoning. In fact, it appears that whenever the volcanic action has forced a new vent upon any fresh point of the earth's surface, the elastic fluids are almost invariably discharged by the same aperture through which the lava is emitted.

When, however, repeated eruptions from the same vent have raised a volcanic mountain of considerable dimensions, the aeriform explosions take place for the most part from its principal or central aperture, which is usually at the summit of the cone; while the lava is rarely discharged from the same orifice, but finds an issue from one or more rents in the side of the cone, or even at the plain at its base. The cause of this is extremely clear. The principal aperture or great crater of the mountain must necessarily be situated immediately above the fundamental volcanic vent, formed by the first explosions of the volcano through pre-existing strata; since the elastic fluids evolved from thence, and tending to rise perpendicularly into a rarer medium, will of course for the most part escape by that aperture. The density of the lava on the contrary opposing the greatest impediment to its elevation to the summit of the mountain, the immense lateral pressure of the ascending column on the internal walls of the passage, or chimney, through which it swells upwards, must, in most cases, aided by the intensity of its heat, force a crevice in these walls, through which it issues into open air, upon the flanks, or at the foot, of the mountain.

Examples of eruptions characterised by these circumstances are abundantly common in the annals of all compound volcanic mountains.

Let us take Ætna for example:

In the eruption of 1536, twelve different mouths opened successively, one below another, on the same line or fissure; each producing lava, while the central crater vomited vapour and scoriæ.

In 1669, the S. E. flank of Ætna is described as having been split open by an enormous fissure, reaching from the summit two-thirds of the distance down the mountain. From its lower extremity issued the vast current of lava, which, taking the direction of Catania, destroyed a third of that town, and formed a large promontory projecting half a mile into the sea.* After the emission of the lava current had ceased, that is, when the internal column of lava had subsided to the level of the lateral vent, aeriform explosions succeeded from the same orifice, and continued to be discharged with violence during fourteen days. The fragmentary matters eructed by them pro-

* Ferrara. Campi Phlegræi. (Borelli. Storia del Eruzoine.)

duced the large cone called Monte Rosso, near Nicolosi, and covered a circuit of about two miles radius, with a deep deposit of black sand containing innumerable separate crystals of augite. This district is only now beginning to support a scanty vegetation, in spite of the assiduous efforts made by the inhabitants to fertilize it. Part of the fissure then formed is still visible behind the Monte Rosso.

In the eruption of 1792, Ferrara observed that a fissure was broken through the side of the mountain; whence, during ten days, the lava boiled out very tranquilly, while the aeriform explosions took place only from the principal crater. At the end of this time the explosions ceased from the main crater, and commenced from the extremity of the fissure, at the same moment that the lava ceased to flow out; the liquid column within the vent having evidently lowered itself, by continual emission, to the level of the lateral aperture. In 1780 the earth sunk along a straight line from the great crater to a new lateral vent which produced the eruption, showing the existence of a fissure in that direction. In 1809, numerous points emitting lava opened successively, upon one line, or fissure, reaching downwards from the margin of the great crater. A similar circumstance occurred during the eruption of Ætna in 1811—12, according to the relation given to me by Signor Gemellaro, who was a witness of the fact. It appears, that after the great crater had by its violent detonations for sometime testified that the ascending lava had nearly reached the summit of the mountain by its central duct, an unusually violent shock was felt, and a stream of lava broke out from the side of the cone, at no great distance from its apex. Shortly after this had ceased to flow, a second stream burst forth at another opening, considerably below the first; then a third still lower, and so on till seven different issues had been thus successively formed, all lying upon the same straight line, prolonged from the summit nearly to the base of the mountain. This line was evidently a perpendicular rent produced in the internal frame-work of the mountain; which rent was probably not produced at one shock, but prolonged successively downwards by the lateral pressure and intense heat of the internal column of lava, as it subsided by gradual discharge through each vent.

The flowing of lava from each of these orifices was followed by the eructation of scoriæ, by which as many small parasitical cones were produced.

In fact, in every lateral eruption of Ætna such a fissure has been formed; the lava issuing from its lower extremity, and successively from different points, as the fissure was prolonged downwards.

Other volcanos present the same phenomena. We may particularize the great eruption of Vesuvius in 1784; when five small cones were produced at the east base of the mountain. They still exist, and mark the points whence the lava torrents issued, by which Torre del Greco was overwhelmed.

In all these cases the central craters of the two volcanos continued to discharge torrents of elastic fluids, carrying up scoriæ, sand, and ashes, while their lavas escaped in currents at a much lower level. When, however, the aeriform explosions took place from the lateral vents, those of the central crater ceased for a time, but recommenced when the former had in turn stopped.

But one of the most remarkable and instructive instances of analogous facts is to be found in the eruption which tormented Iceland in 1783, when the lava issued in enormous quantity from three sources opened successively in a plain at the foot of the high volcanic cone, (Skapta Jocul) from which the gaseous explosions had been, and for some time continued to be, discharged.

These scources were about eight miles distant from one another, and were formed along the same straight line, which obviously marked the direction of a fissure created in the superincumbent strata of the plain, by the upward pressure of the lava below, in communication with that which was forced up the internal chimney of the neighbouring mountain. A fourth source seems to have opened itself in the prolongation of the same line, but beneath the sea, and at a distance of 30 miles, producing a rocky island, now reduced to a mere shoal by the erosive action of the waves and submarine currents. The lava produced by the three terrestrial sources deluged the plain to a superficial extent of more than 40 miles by 30.

In this instance it appears that the frame-work of the volcanic mountain offered a more solid texture and a greater degree of resistance to the pressure of the internal column of lava, than the superincumbent beds constituting the plain at its base, which in consequence were the first to give way, and open an issue to the fluid.

The immense quantity of lava produced, and the velocity with which it flowed forth, were obviously in direct proportion to the great height to which the column had ascended in the interior of the mountain, before the fissure was formed. The distance of the apertures by which the lava was poured out. from the crater or central vent of the volcano, which discharged the aeriform fluids, proves the vast horizontal extent of the subterranean reservoir of lava; which is the less astonishing from our knowledge that the whole island of Iceland has been produced from the bottom of the sea by the successive eruptions of the same volcanic system, we might almost say, of the same *volcano.*

The gigantic volcanos of S. America appear to have frequently presented examples of eruptions taking place under similar circumstances, but the accounts of which we are possessed are not sufficiently detailed for the purpose of comparison.

It may very plausibly be suspected that each of the shocks by which the environs of a volcano are so repeatedly agitated, during, and previous to an eruption, is occasioned by the rending of some part of the solid frame-work of the mountain or its supporting strata, by the action of the force we have described as resulting from the pressure in all directions of the liquid which is in communication with that elevated within the volcanic chimney. The prolongation or widening of a fissure previously formed would have the same jarring, or vibratory effect, as the creation of a new one. It is indeed a remark common to the observations made on almost all volcanic eruptions, that local earthquakes always precede the emission of lava currents, and cease while the lava is flowing, to recommence when it has stopped. Those shocks, on the other hand, which are felt to a considerable distance, are probably caused by new rents produced in the solid subjacent strata supporting or surrounding the mountain, and enter into that class of earthquakes which were discussed in a former chapter.

The rents produced in the frame of a volcanic mountain by the disjunctive force in question, are sometimes of such width that the whole mountain appears cleft in two, as occurred to the volcano of Machian, one of the Moluccas, in 1646. The crater of the Souffrière of Montserrat, and the volcanic cone of Guadaloupe, both appear to have been thus split through by a large fissure. The Montagne Pèlée of Martinique offers a similar fact, (Moreau de Jonnes.) Even the eruption of Vesuvius in October, 1822, which was peculiarly fertile in interesting phenomena, offers also an example of this rending of the mountain. The crater or rather chasm left by that eruption is itself an enormous fissure broken across the cone in a direction N. W.—S. E. It was prolonged through the whole frame of the cone on the S. E. side, and produced a deep notch in the mountain, which though considerably effaced by the beds of scoriæ and fragments thrown into it, is still 500 feet lower than the neighbouring point of the ridge of the actual crater.

§. 3. The fissures thus formed in the frame-work of the mountain and instantly occupied by the liquid lava, become hermetically sealed, by its subsequent consolidation, and assume the character of *dykes*. These being usually formed, as has been said, in a vertical direction, and therefore transversely to the lateral beds or currents of lava, communicate a vast accession of strength to the structure of the mountain, acting as ties to the latter, which may be likened to the main beams of the edifice.

The section of Monte Somma presented by the steep cliffs above the Atrio del Cavallo, which are the remaining walls of the ancient and principal crater of this volcano, exhibits an infinity of such dykes traversing the mass of the mountain in numerous directions, all approaching, however, more or less to the vertical, and crossing each other, and the more massive and nearly horizontal beds of lava and scoriæ, so as to give a reticulated appearance to the rocks. They are of very compact basalt, and frequently divided into prisms lying at right angles to the walls of the dyke. The recent crater of Vesuvius (1823) at present exhibits similar features, which are common in fact to all volcanic mountains of which the internal structure is sufficiently exposed.

From this consolidation of the matter occupying the rents of a volcanic mountain, it happens that at every fresh effort of the expansive force a rent must be forced in a new place. In fact, there is no instance extant of an old lateral aperture having opened a second time.

§. 4. Thus it is clear that the augmentation of the cause that tends to destroy the fabric of the mountain, viz. the height of the internal column of lava, is compensated by a parallel increase in its strength; and no other limit would appear to be set to the increase of a volcanic mountain, in height as well as bulk, than the cessation of the phenomena which erected the whole, the absolute extinction of its outward activity.

In fact, we are acquainted with examples of such mountains reaching to an extraordinary altitude. Ætna, the Peak of Teneriffe, the volcanos of Kamskatchka, of the Aleutian Isles; those of the Andes and Mexico, which a short time back were supposed the highest on the globe, are of this nature, and have been apparently raised in the mode above described by the continued accumulation of loose ejected matters fortified by lava currents and dykes. It is true the instances of the Pic de la Teyde, Chimboraço, and some other very elevated volcanic mountains which have not for some centuries exhibited eruptions from their summits, would appear at first sight to show that another limit exists, and that when the mountain has acquired a certain height it becomes unable to resist the pressure of so enormous a column of lava, which in consequence invariably finds a vent at its side or foot.

But it must be recollected that these lateral eruptions tend continually to fortify the sides of the mountain by the accumulation and consolidation of their products; and, it may be foretold, therefore, that should these volcanos continue in activity, the time will come when they will be sufficiently propped on all sides, to be able to sustain the pressure of the lava; which, in this case, will be again poured forth from the summit.

§. 5. These lateral eruptions proceeding exactly in the manner of those from fresh vents described above (chap. iii.) like them produce a more or less regular cone on every point where they find an issue.

The slopes of Ætna are loaded with above 70 such *parasitic* cones; many of them of considerable magnitude,* each has its crater, and each marks the source of a current of lava; that from the Monte Rosso destroyed Catania in 1669.

Vesuvius has occasionally produced similar hills—that on which stands the Camaldoli della Torre is an example, and eastward of this point rise five other small neighbouring cones thrown up by the eruption of 1760. The lava from these apertures overwhelmed Torre del Greco.

But the great cone of Vesuvius was itself in fact originally but a parasitic cone thrown up in the centre of the old crater of Somma, of which the remaining circumference completely embraces Vesuvius from east to west by south.

Exactly in the same manner has the cone of the island of Vulcano, which contains the large crater now in the state of a solfatara, been thrown up in the centre of a still larger and more ancient crater, which still incloses it by a semicircular range of cliffs, composed of alternating beds of lava and ashes. See the accompanying plan of the island of Vulcano, and also its profile in Plate II.

* The Monte Rosso was estimated by Hamilton to be *half a mile high* from its base, and three miles in circumference.
It is probably about 500 feet in height, and two miles round.

Fig. 24.

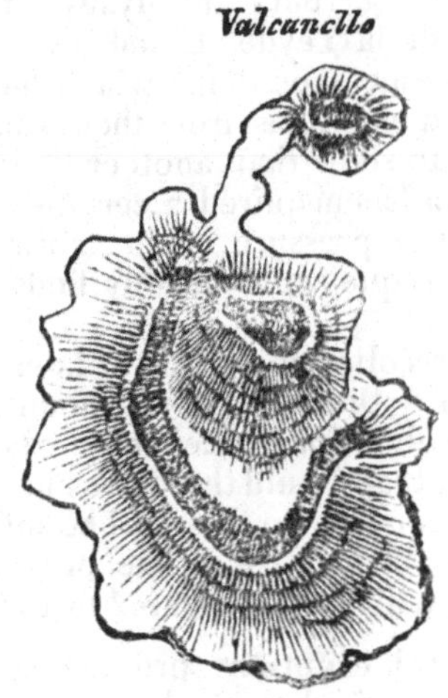

Island of Vulcano.

§. 6. Of the fragmentary matters ejected, whether by the central or by lateral vents of a volcanic mountain, the lightest and most pulverized particles are scattered by the winds, often to a considerable distance.

The fragments of intermediate size and weight are carried down the slopes of the mountain by aqueous torrents. These have their origin in the quantity of rain which as has been noticed above, generally falls during every volcanic eruption, the consequence of the vast volumes of aqueous vapour evolved, and to which every such mountain is the more exposed in proportion to its height. If its elevation, or geographical position, cause it to retain frozen snows upon its sides or summit, still more sudden and terrific deluges of water are created by the showers of red-hot scoriæ which fall upon them, or the contact of the still hotter lava.

In Iceland phenomena of the latter kind accompany almost every eruption from its snow-clad heights, and constitute by far the most destructive part of the volcanic catastrophes, to which the inhabitants of this immense sea-girt cauldron are exposed.*

* During the eruption of Katlegiaa in 1756, torrents of ice, rocks, and sand, occasioned by the melting of the glaciers, produced three parallel promontories, reaching three leagues into the sea; which remain above its level in places where the fishermen formerly found 40 fathoms of water. (Olafsœn and Povelsen.)

The inhabitants of the slopes of Vesuvius designate these torrents of mud and scoriæ, by the name *Lave d'acqua*, in contra-distinction to the lava currents or *Lave di fuoco*.

Thus, alluvial deposits of fragmentary matter may be expected to fill up the hollows, cover the skirts, and extend to some distance from the base of every volcanic mountain, and often to alternate with those of its lava currents, which from their superior fluidity have extended farthest from the central summits. And such appears to be the structure of these mountains wherever they have been geologically examined. (Mt. Dor, Cantal, Mezen, Teneriffe, Hungary, &c.)

The trees and plants growing on the slopes of a volcanic mountain must be often rooted up and carried away by these aqueous debacles, and buried amongst the alluvial strata at its foot.

This is no doubt the origin of the fossil wood frequently met with in the Tufas of the Mont Dor and Cantal, of Hungary, and Iceland, (Surturbrand.) This is occasionally mineralized, but not always. In Hungary it is sometimes silicified, and changed into opal. In the Mont Dor, I have seen the trunk of a tree, which at one extremity was reduced to perfect jet, and at the other retained the colour and texture of wood, and was not at all carbonized.

These alluvial matters often acquire on desiccation a great degree of solidity, even without being cemented by ferruginous or other infiltrations. The ashes of the eruption of Vesuvius in 1822, carried down the slopes of the hill by torrents of rain, caked into a hard rock, which required a sharp blow from a hammer to break it.

§. 7. If the sea, or any inland lake, washes the base of the mountain, such fragmentary matters as are carried down into it by torrents, or fall there by the mere force of projection, may be distributed on the bottom or shores of the ocean by its currents, and mixed or interstratified with its other more purely marine deposits.

This appears to have occurred in the stratified Tufas of the Terra di Lavoro, and the campagna di Roma. In Hungary pumice conglomerates alternate with tertiary limestone strata, &c.*

In the neighbourhood of Pont du Chateau, (Auvergne,) a calcareous peperino alternates with the strata of limestone

* It is by no means necessary to conclude, as M. Beudant has done, from this, that the trachytic volcanos of that country were *submarine*. It is clearly sufficient that they were insular, or washed at their bases by the sea, to produce this result. *Voyage en Hongrie*, iii. 570.

containing numerous fresh-water shells; and on Gergovia, in the same neighbourhood, currents of basaltic lava are interbedded with fresh-water limestone.

The accompanying figure presents an ideal section of a volcanic mountain after one of its paroxysmal eruptions.

Fig. 25.

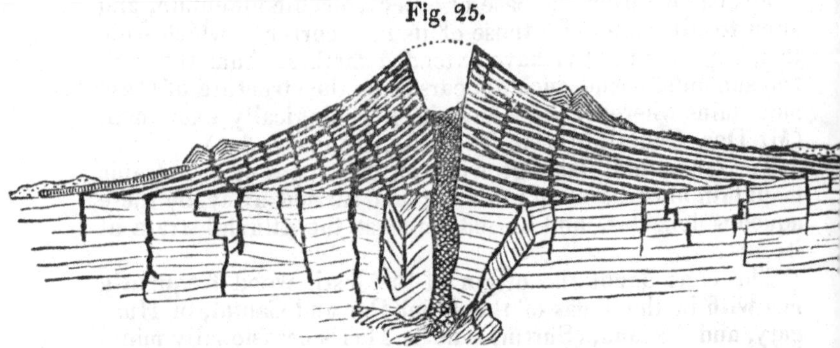

§. 8 In the case of simple cones, notice has already been taken of the figure and origin of their *craters*, or the cavities included in their interior slopes, and having the orifice of eruption at the bottom. Those cones or volcanic mountains which are the produce not of one but repeated eruptions from the same vent, usually possess a similar central cavity, which however is subject to various changes and modifications, and occasionally altogether disappears. These changes, it will be seen, are the necessary consequence of the repetition of successive eruptions from the same orifice; for

I. When, in any individual eruption, the force with which the fragmentary ejections are launched into the air, is inferior to that of the preceding eruption, these matters will all fall within the circumference of the crater, formed at that time, and tend to fill it up by the production of a parasitical cone within this enclosure.

II. The lava also, swelling up through the volcanic orifice, must occupy the bottom of the crater; and if its weight does not succeed in forcing a passage through the side of the mountain, it will fill up this cavity to a height proportioned to its volume.

The repetition of these circumstances always accompanying every minor eruption, tends to fill up the whole cavity of the crater produced by the last violent or paroxysmal eruption, until the matters elevated, both liquid and fragmentary, find their way at length over the ridge of this basin. By this suite of operations the crater of a large volcanic mountain becomes temporarily replaced by an irregular plain, at length possessed of a degree of convexity, and often rendered still more unequal by the small parasitical cones which every minor eruption throws up on its surface.

In this state of things the external repressive forces which consist in the weight and resistance of the matters obstructing the habitual vent of their volcanic energy are at the maximum; the products of subsequent minor eruptions escaping down the outer slopes of the cone.

If after this moment any violent eruptive paroxysm take place, by the creation of a deep fissure penetrating into an intensely heated point of the volcanic focus,—the rapid, powerful, and tremendous explosions that ensue, soon break up and shatter the mass both of solid and fragmentary rocks which had accumulated within the interior of the ancient crater, and project them in fragments to a vast height into the air, whence a part falls on the outer slopes of the mountain, while the remainder which fall back towards the orifice of projection are again and again driven forth till they are triturated to the finest possible degree of comminution. The final result of this process is, to leave a crater bearing marks of the violence with which it has been torn through the bowels of the mountain, and of a diameter and depth proportioned to the energy of the eruption.

The encircling walls of such a cavity, will be more or less perpendicular, and present sections of the beds both of consolidated lava and conglomerate, through which the aeriform explosions forced their way. In process of time these walls will crumble in, and their fragments, forming a talus at the bottom, will soften off the abrupt features of this cavity, and diminish the steepness of its enclosure.

The crater of a volcanic mountain, as well as of a simple cone, usually deviates considerably from the circular form, being lengthened in the direction of the fissure from the enlargement of which it proceeds.

The phenomena of Vesuvius, during the last century, will serve to exemplify the series of changes we described as incidental to the principal or central crater of an habitual volcano.

In 1766 the great crater, which previously existed, is described by Sir W. Hamilton as entirely choked up, and in the place of a cavity he found an irregular circular plain at the summit of the cone.

This plain was subsequently broken up, and the cone completely *gutted*, by the great paroxysmal eruption of 1794, which left a vast cavity, described as of tremendous depth and width, and surrounded by perpendicular crags. The process of replenition then again commenced, till this crater was once more so completely choked by the accumulation of matter produced by minor eruptions from its bottom, that during the years 1813—1822, a rude plain of the same character as that described by Hamilton, again existed at the summit of the cone. The plain towards the latter part of this period had acquired a convexity of form, and supported many minor parasitical cones. This state of things continued until, in October 1822, another eruptive paroxysm occurred, similar to that of 1794; which, breaking up the plain and launching its fragments into the air, reproduced a central cavity, or crater, of vast size and depth, and of an irregular elliptical form, being obviously the result of the forcible enlargement, by rapid and violent explosions of steam from below, of a fissure broken through the framework of the cone in the direction N. W. by S. E.

The great crater of Ætna is, in its actual state, an example of one in which the process of replenition has already made considerable progress. When I visited it in 1819, there existed a small parasitical cone in full activity at its bottom, whose frequent eruptions must tend materially to lessen the depth of this cavity.

Hence it is observable, that the principal or central crater of every volcanic mountain, or cone produced by *numerous eruptions from an habitual vent*, is the result of the last paroxysmal eruption of the volcano.

Thus too it appears, that besides the operation of time and external causes in damaging or totally destroying the figure of craters, there exists in the volcanic phenomena themselves a tendency to obliterate their forms—a series of causes alternately scooping out the interior of a volcanic mountain, and gradually filling up again the cavity, or crater, thus produced.

The neglect of this observation has been the source of numerous errors amongst those Geologists who have turned their attention to volcanic districts; many of whom have frequently overlooked or rejected some of the most decidedly marked craters, from their want of perfect symmetry; while others have drawn negative conclusions as to the volcanic nature of a mountain, from the absence or obliteration of its crater, which is in fact as regular and necessary a result of the volcanic phenomena as their formation.*

It must be observed, that the explosions by which a crater is hollowed out from the centre of a volcanic mountain, are chiefly those which take place subsequently to the emission of the lava in a liquid form, and by which the eruption terminates.

This consideration would have prevented Dolomieu from the error of supposing that the large crater of Volcano (one of the Lipari isles) must have been filled to the brim with liquid glass before the small current of obsidian which flowed from its summit in 1786, could have been produced. The emission of this lava was here, as in most similar cases, antecedent to the excavation of the crater.†

§. 9. The dimensions of the cavity formed in this manner will be determined by the violence of the eruption, and the more or less yielding nature of the walls of the eruptive fissure. When these circumstances are favourable in the extreme, the terrific explosions that ensue, and which forcibly enlarge the crevice that propagates the expansive process to the intensely heated lava below, may not only break up and drive into the air the matters which have during a long previous period accumulated in the old crater, and obstructed this habitual vent of the volcanic heat, but even tear up their way through, and destroy, the greater part or the whole of the volcanic mountain itself, leaving in its place a wide cavity encircled by the ruins of the shattered cone. Examples of such an occurrence have been met with even in present times,‡

---

* In Hamilton's Campi Phlegræi are drawings of the variations of form of the cone of Vesuvius, from the years 1766 to 1780, which illustrate the progressive elevation of a minor cone in the interior of a crater, until this basin is gradually filled and obliterated.

† For an ideal section of a volcanic mountain after a paroxysmal eruption, see page 83, fig. 13.

‡ Such a catastrophe destroyed, in the year 1638, a colossal cone called the Peak, in the Isle of Timor, one of the Moluccas. The whole mountain, which was before this continually active, and so high that its light was visible, it is said, three hundred miles off, was blown up and replaced by a concavity now containing a lake.

The isle of Sorea, another of the Moluccas, disappeared entirely in a similar manner, during a violent paroxysmal eruption in 1693.

Again, according to M. Moreau de Jonnes, in 1718, on the 6—7 March, at St. Vincent's, one of the Leeward Isles, the shock of a terrific earthquake was felt, and clouds of ashes were driven into the air with violent and in the remains of ancient volcanos now apparently extinct, or at least dormant, we may observe proofs of these phenomena having not unfrequently taken place.

Explosions of this extreme violence are naturally to be expected to characterise the return of a volcano to a condition of external activity, after a long period of apparent repose occasioned by the long predominance of the repressive forces. And such a burst seems to have taken place in the great eruption of Vesuvius, A. D. 79, by which half of the old cone of the pre-existing mountain was blown into the air, and buried Herculaneum, Pompeia, and Stabiæ, under its triturated ruins. The remaining half of the cone exists in the Monte Somma, and there can be little doubt that the modern cone of Vesuvius owes its origin to that eruption.

The still active volcano of the Isle of Bourbon presents a remarkable analogy of situation to Vesuvius, rising in a similar manner from the centre of a vast semi-circular enclosure, formed by precipitous cliffs, which consist, like those of Somma, of alternate beds of basalt and scoriæ, whose inclination and direction prove them to have been produced by a volcanic vent, nearly in the same position as the present one. In this case, as well as that of Vesuvius, the appearances are only to be explained by supposing half of the cone of an ancient volcano existing on the spot to have been suddenly blown up by a severe paroxysmal eruption of the nature of those we have supposed above, which had been preceded by a longer than ordinary interval of outward inactivity.

This paroxysm seems to have been succeeded by the lengthened phase of moderate and prolonged activity which has characterised this interesting volcano from the period at which the island was first colonized unto the present time. Indeed these circumstances are common to many volcanic mountains. The island of Stromboli presents a cone rising from within the circuit of an older crater. De Buch observed, in the island of Teneriffe, a semi-circular wall of precipitous rocks 2500 feet in height on some points, and of immense diameter, which partly encloses within its limits both the cone of the Peak and that of Cahorra, almost the rival of the former in magnitude. This enclosure is evidently the remains of a vast crater or chasm formed at some

---

detonations from a mountain situated at the eastern end of the island. When the eruption had ceased it was found that the whole mountain had *disappeared*. A hurricane accompanied this catastrophe, caused, it may be presumed, by the ascending draught of air created by the explosions.

distant epoch, when the upper half of this volcanic island was blown into the air by one intense eruptive paroxysm.

The volcano of Pichinca, in Mexico, visited and described by Humboldt, has an elliptical crater of this nature 1400 yards in diameter, and between four and five thousand feet in depth : its inner sides consist of steep and tremendous precipices. Many parasitical cones are observable at the bottom; one of which at least is in activity, since the red-hot scoriæ it projects are visible by night.

The circular hollow called Astroni, in the Campi Phlegræi, near Naples, as well as the Solfatara, may be mentioned as additional examples of craters produced in this manner. The former exhibits, in the middle of the plain which forms its bottom, a small parasitical cone, the result of a minor eruption subsequent to the formation of the enclosing cavity. Precisely the same fact is observable in the basin of the lake of Roneiglione, a much larger specimen of such a crater, from the centre of which rises a secondary cone of considerable dimensions, itself possessing a very regular crater. This last also is one out of many examples of the occupation of these craters by a lake; a circumstance which necessarily results from their form, whenever the conglomerates which line the interior slopes of the basin are not permeable to water.

It is indeed highly probable that all those remarkable circular lakes which are so commonly met with in volcanic districts, and particularly in the west of Italy, central France, and the left bank of the Rhine, owe their origin to the destruction of the upper and central part of a volcanic cone by a paroxysmal eruption.

The skirts, or talus, of the mountain would, after such a catastrophe, remain as an enclosure to the cavity if produced, and in most cases present an abrupt precipitous escarpment towards the interior, and a gradual and shelving slope outwardly. Such are, in reality, the characteristic features of almost all these natural basins.

Their forms have been of course, in some degree, smoothed down and softened by the accumulation of fragmentary matter ejected by the paroxysmal eruption, and occasionally by those of subsequent explosions of a minor nature, which, as was observed with respect to the Lagodi Roneiglione, have sometimes raised up parasitical cones within their interior.

Thus, where the volcanic mountain consisted of a more than ordinary proportion of incoherent matters, and those of a light and loose texture, such as lapillo and felspathic tufas, the basin is encircled by ridgy elevations composed of these substances, to which the influence of time and meteoric denudation has since given a still gentler outline.

As examples of such basins may be instanced, those of Agnano Averno, and Roneiglione in Italy; of Laach, on the left bank of the Rhine; the Pulver-Maar of Gillenfeld; the Ulmener, Dreiser, and Mosbrucher-Maare, in the Eiffel. In the island of San Miguel is a crater of this kind occupied by two lakes, and fifteen miles in diameter.*

§. 10. But all the circular lake basins of volcanic districts do not result from the destruction of a volcanic mountain and the enlargement of its central crater. Many such are to be met with, which have been evidently produced by a single eruption from a new vent, through rocks previously intact; consisting of a circular elliptical cavity hollowed out of these rocks, whether primitive or secondary, and surrounded by a greater or less accumulation of fragmentary ejections. In fact, if we suppose the focus of an eruption from any fresh point to lie more than ordinarily deep; while, at the same time, its expansive force is extreme, the explosions of steam which discharge themselves with vast rapidity through the fissure of communication with the outer air, will enlarge the point of this fissure, on which they escape into an opening of very considerable diameter; breaking up and shattering into fragments the rocks on either side; while, at the same time, from the depth of the fissure, the lava will be unable to rise to its mouth, and only a small proportion of this substance will be discharged outwardly in the form of scoriæ; the weight of the lava, elevated in the vent, alone sufficing to stifle the focal ebullition, and prevent any farther intumescence. The result of such an eruption would be the formation of exactly such a crater or cavity as we have described above; and examples of which are to be seen in the Lacs de Thalaba Guèry and Paven (Auvergne); in the Meerfelder Maar, and more than twenty other similar lakes in the Eiffel.

The very frequent occurrence of these basins in the Eiffel, where they have been opened through a formation of greywackè slate, confirms the above theory of their origin; not only because of the comparatively friable nature of this rock, but also since, in this instance, the focus may be reasonably

---

* Webster's Account of the Island of San Miguel.—Boston, 1818.

conjectured to lie deeper than where the primitive rocks show themselves on the surface.

Among the lakes of this kind quoted as existing in France, that of Thalaba is excavated in granite; the others in the basaltic or trachytic beds of a vast volcanic mountain. The latter are of course to be considered in the light of subsidiary apertures subordinate to the great central crater of the volcano.

The lake of Gustavila in Mexico, described by Humboldt, and of which an engraving is published in his "Vues des Cordillerès," is a remarkable example of the same kind.

§. 11. This leads to the consideration of another cause destructive of the regular form of volcanic cones, which has not as yet been alluded to.

This consists in the bursting of lakes which have formed themselves within their craters. Whenever the pulverulent ejections of a volcano are of such a nature as to form a clay or paste on mixture with water (and this is the case with the felspathic lavas in general), it is obvious that the quantity of these ashes, which always accumulate at the bottom of a volcanic crater, mixing with rain water, must soon form an impervious coating to the interior of this basin, and force the water falling from the skies to collect into a lake. This body of water, under favourable circumstances, will increase gradually in depth and volume, until it finally overcomes, on some point, the resistance offered by the loose and fragmentary matters forming the sides of the cone; breaks through this inefficient barrier, and rushes downwards to the lower levels, carrying with it the mass of loose substances through which the torrent has forced a channel, and spreading them over the surface on which it discharges itself. Thus producing alluvial, or rather *eluvial*, formations of considerable volume.

The bursting of such lakes may also be occasioned by other more immediate causes than the mere weight and lateral pressure of their waters; viz. by the occurrence of an earthquake, or a fresh eruption from the volcano beneath; which in this case will give rise to what are vulgarly called *mud eruptions*, and igneous eruptions, at the same time.

These aqueous or *eluvial* eruptions appear to play a great part among the phenomena of the South American volcanos. Torrents of mud or fine volcanic ashes mixed into a paste with water, occasionally burst from the sides or summit of these stupendous volcanic mountains, carrying destruction into the vallies and plains at their base. (Moya of Quito.)

Vast numbers of fish are occasionally found enclosed in the

matter of these torrents. These have obviously been the inhabitants of the crateral lakes, the bursting of which occasions those deluges; which are always preceded, and of course caused, by violent earthquakes.

The carbonaceous matter which so remarkably characterises the substance of these mud eruptions, is probably owing to the Algæ or other water plants that grew within the lake.

This explanation of the origin of such phenomena is confirmed by the fact, that they only take place amongst the trachytic volcanos, the triturated particles of the basaltic lavas seldom mixing into a retentive paste with water.

Amongst the volcanos of the European continent, such phenomena have formerly occurred; although, at present, from the non-existence of trachytic volcanos in a state of activity, they are not observable.

To such circumstances are obviously owing the trass of the Andernach district, of the Eiffel, and Siebengeberge; and many of the tufas and breccias of the Mont Dor, Cantal, and Mezen; of the Campi Phlegræi, and Ischia; and in all probability of Hungary.

The variety of these fragmentary eluvial deposits originates, of course, in the different mineral characters of the comminuted lavas, and their being, in consequence, more or less permeable to water, more or less apt to *set* or cohere together on reaggregation, as the water gradually drained off from them, &c.

The contraction which they experienced in the process of desiccation, joined to the influence of the forces of attraction, has, in many instances, given them a divisionary structure, occasioned, no doubt, by exactly the same process, and regulated by the same laws, which we have traced above in the case of lavas consolidated by the condensation or exudation of their fluid vehicle; productive also of the same modifications of form, viz. the tabular, lamellar, globiform, prismatic, and columnar, in all their variety.

The divisionary structure in these conglomerate rocks is, however, less frequently observable, and when it occurs less perfect and remarkable than in lithoidal lavas.

When fissures of any width were formed in these conglomerates, during the process of desiccation, they have been filled by exudation of the finest matter remaining fluid in the vicinity of the crevice.*

* Veins of a very fine-grained tufa vertically cutting through beds of a coarser grain and porous texture, and to which this origin alone can be

These conglomerates, like those produced by ordinary alluvia, are found of course to contain occasionally organic remains, such as trunks of trees growing on the slopes of the mountain at the time of its disruption, or the bones of animals which were destroyed by this catastrophe.* The trass of the Rhine district is remarkable for enveloping a great quantity of vegetable remains; they are always in a carbonized state, and occasionally are found to contain passages into jet. The Surturbrand of Iceland was no doubt deposited under somewhat similar circumstances.

§. 12. The minor eruptions of an habitual volcano, particularly when in the first or second phase, will usually take place from foci seated in the axis, or what may be called the chimney, of the mountain, and above the level of the fundamental vent, originally forced through the superficial strata by the first explosions of the volcano, and through which all the materials of the mountain have been successively protruded. These secondary phenomena being caused by the expansive force of the heated, but solidified lava, elevated into this position.

The paroxysmal eruptions on the contrary must be supposed to proceed from the penetration of some deeper cleft into the more intensely heated lava mass below this vent, occasioned by the general expansive efforts of this subterranean bed.

§. 13. So long as a volcanic mountain preserves a sufficiently conical form, it will be obvious that the eruptions have remained constant to the same fundamental vent.

Occasionally, however, the ancient and central crater of a volcano appears deserted or "extinct;" the aeriform explosions habitually escaping by some new aperture, at a greater or less distance from it on the flank or at the foot of the mountain. In this case the mountain loses its regularity of form, a new cone gradually accumulating around the actual orifice of eruption.

---

attributed, are observable in the formations of the Campi Phlegræi, and particularly at Capo di Monte.

* Thus, after the great eruption of Tomboro in Java, in 1818, the sea in the vicinity of the island is described by eye-witnesses as covered by floating pumice and half-burnt trunks of trees, torn of course from the shattered sides of the mountain, and launched with its fragments into the air, by this great paroxysmal eruption.—Journal of Science, vol. i. p. 255.

There is then reason to suspect that the original fundamental vent has been permanently obstructed by the immense weight and cohesion of the matters accumulated above and within it; and that a new vent has been produced through the pre-existing rocks, corresponding to the new and habitually active aperture.

The island of Bourbon offers an example of such a change. The vast central crater occupying the peak of the immense submarine mountain which constitute the greater part of this island, has been to all appearance completely inactive for a very long period. Since its extinction the habitual eruptions of the volcano have apparently twice shifted their position; once to a distance of about 15 miles to the S. E. in the Plain des Calaos, and a second time to point at a much less distance, but on the prolongation of the same line or subterranean fissure. At the first of these two spots a mountain of considerable magnitude has been formed, though inferior to the principal eminence. At the latter, the cone of the actual volcano has been thrown up, and continues rapidly increasing in height and bulk, from the frequent, and almost permanent, eruptions, which take place from its summit.

When the principal crater of an habitual volcano has been in this manner deserted, it is often reduced to the condition of a Solfatara, by the continued emanation of acid vapours from the mass of heated but solidified lava remaining in the chimney of the mountain. These vapours force their way by percolation through the pores of the rock or crevices too narrow to permit any fresh intumescence below, and by the gradual loss of caloric thus sustained the residuum of lava is eventually cooled down. The process may, however, last for centuries, as has been the case with the Solfatara of Pozzuoli, which from the white colour of the altered lava-rocks was called Λευκογεος by Homer.

These slow emanations of vapour appear to be almost entirely confined to the trachytic volcanos of which the lava, as has been remarked in a former chapter, when solidified, is usually more porous than basalt; and this is a strong confirmation of the correctness of the origin here attributed to them. On a former occasion instances of Solfatare have been given, and the process of alteration which the lava and conglomerate rocks undergo by the percolation of acid and heated vapours through their pores or crevices, described. The various saline and metallic depositions, which result from these changes, have also been already enumerated.

CHAP. VIII.

Subaqueous Volcanos.

---

§. 1. As yet our attention has been confined to the phenomena of those volcanic vents which open at once in the atmosphere; but it must not be forgotten that such apertures are liable to be created on any point of the globe's surface, and therefore, as well on those which are covered by permanent bodies of water, as on the dry surfaces of our continents.

Indeed, from the far greater extent of surface of the former character, which exceeds that of land in the proportion of three to one, we might expect the number of subaqueous eruptions infinitely to exceed those that take place in open air. It must, however, be remembered, that the repetition of eruptions from the same vent has the effect of raising the apex of the submarine volcanic mountain above the level of the sea; and that, with the greater number of habitual submarine vents, this limit of elevation has already been reached, and they have become visible as insular volcanos.

It is necessarily very seldom that an opportunity is afforded of observing a volcanic eruption from the bottom of the ocean, or any inland sea. The resistance created by the greater density of the medium must, as we shall shortly see, prevent the explosions, in most instances, from reaching its surface, and at all times from attaining any great height above that level; and, consequently, from being visible at any considerable distance; and it is only to the crews of vessels casually passing, that such an opportunity can occur. We must not expect, therefore, to be possessed of many accounts of such phenomena.

§. 2. Instances, however, are not wholly wanting, and the details of what has been observed on these occasions, lead us to conclude that the volcanic phenomena take place exactly in the same manner from the bottom of the sea, as from the open surface of a continent, subject only to the modifications produced by the greater density of the surrounding medium, and the greater external pressure, caused by the weight of the overlying column of water, which in this case becomes one of the elements of the repressive force. In fact, it cannot be otherwise, since we know that the instant the cone of the submarine volcano raises its peak above the waves, it enters into the class of subaerial volcanos, and the nature of its activity is, in all respects, precisely the same as that of a continental vent.

The instances, which have been witnessed, of submarine eruptions, are;

1. An eruption off St. Michael one of the Azores in 1638,* other eruptions took place from the spot in 1691 and 1720, which produced an island six miles in circumference.

2. That which gave birth to the Isola Nuova off Santorini in the Grecian Archipelago in 1707.† Santorini itself is reported by Pliny to have been produced in the same manner B C 236, as well as two other neighbouring Islets, Jera and Thia. These were enlarged by subsequent eruptions.

3. That of Sabrina, near St. Michael, one of the Azores, in 1811.

4. An island thrown up at some distance from the coast of Iceland during the violent eruption from Skapta Jokul in 1782.

5. A new volcanic island formed in the Aleutian groupe near Uanalashka in the spring of 1814, called Bojuslaw by the Russian hunters.*

On all these occasions columns of smoke by day, and flames (jets of red hot scoriæ) by night, were seen to rise from the sea, which was considerably agitated, discolored, and heated to such a degree as to kill numbers of fish.

---

* Sanderson's Hist. of Charles I. Memoires del 'Academie, 1721.
† Hist. de l'Academie, 1708.
* Kotzebue.

At length the dark coloured rocks shewed themselves above the sea-level. These, in the case of Santorini, were both fragmentary and lithoidal (lava), and the island produced by that eruption has consequently remained firm;—in the other instances the fragmentary cone alone appears to have risen above the water-level, and the action of the waves and marine currents upon those loose matters, soon mined and degraded them, reducing the island to a submarine shoal.

The result of actual observation in these instances, and of the whole mass of reasoning à priori, is, that a volcanic eruption taking place from a vent opened below the level of a body of water, conducts itself in a manner very little if at all different, from a subaerial eruption. The same explosions of aeriform fluids are observed;—rocky fragments, ignited scoriæ, and their comminuted ashes, are thrown upwards; the heavier in falling must accumulate round the vent, into a cone with a central crater, while the lighter will be borne to a distance by the tides and currents of the sea. Lava issues, and spreads over the subaqueous bottom, seeking the lowest levels, or accumulating upon itself, according to its liquidity, volume, and rapidity of congelation; following in short the same laws as when flowing in open air.

Humboldt and De Buch have both published as their opinion, that in submarine eruptions the strata previously forming the bottom of the sea, are uniformly elevated, and that positive eruptions do not take place from the vent until these strata have been raised above the surface level of the sea. There is, however, no reason à priori for supposing such an anomalous distinction between the mode of action of subaqueous and subaerial volcanos; and still less is the supposition warranted by observation of such phenomena, since, in all the cases already mentioned, the only rocks which showed themselves above the sea-level were uniformly either lithoidal or conglomerate lavas, the products of the eruption by which they were raised.

§. 3. It is true that the vast augmentation of the repressive force occasioned by the pressure of the column of water above the vent must proportionately impede the ebullition of the lava to which the fissure of escape penetrates; but the resistance to elevation of the solid strata, which form the pavement of the ocean, is augmented exactly in the same proportion; so that these are not more likely to give way to the effects of the expansive force, than when exposed to the pressure of the atmosphere alone. The consequence of this vast increase of the repressive forces will, however, be

1st. That, when the expansive force of the confined lava has at length overcome its antagonist, the tension of the mass, and its temperature, must be proportionately intense; and

2dly. That the vapour which escapes from the lava, owing to its excessive tension, will be speedily refrigerated, by contact with the colder strata of water through which it passes in its ascent, and *condensed* by the immense pressure; so that, unless when the eruption takes place where the water is shallow, *no volumes of vapour will rise to the surface*, and consequently no aeriform discharges will announce the occurrence of the eruption in the depth of the ocean.

Torrents of lava will be produced, and scoriæ and other fragmentary matters thrown up to a certain height, by every such eruption; but until their accumulation shall have raised the orifice of the mountain to such an elevation, that the aqueous vapours it discharges are no longer wholly condensed by the pressure and refrigerating action of the surrounding water as they rise, no other appearance of its eruptions will be visible from without, than what consists in the partial discoloration, and heating, of the water above the volcano.

When therefore the eruption becomes visible, and the vapours produced below are discharged from the surface of the sea, and particularly when they are seen to throw up scoriæ above its level, it may be safely argued that the summit of the submarine volcano is at no great depth, and that few more eruptions are required to raise it above the sea-level, and create a new island, or insulated sub-aerial volcanic mountain.

§. 4. The mineral or saline compounds which, in greater or less quantities, always accompany the aqueous vapour evolved from a volcanic vent, will on the condensation of this vapour mingle with the waters of the ocean, and add to the ingredients of the same nature with which it is impregnated.

Disposition of Products.

§. 5. Though the volcanic phenomena seem to be essentially the same, whether acting from a subaqueous vent, or one in open air, yet the difference of the medium into which the eruption breaks forth, must considerably modify the *disposition* of the substances produced.

Fragmentary Rocks.

A large portion of the fragments thrown up will accumulate immediately round the vent, into a cone; but those which are scattered to any height by the gaseous explosions, and particularly the lightest and finest of the fragmentary matters, will be suspended for a considerable time in the agitated fluid, imparting to it a turbid colour, and upon the cessation of the disturbing causes will be gradually and evenly deposited, over a large surface of the bottom of the ocean, in the form of sedimentary strata.

Hence the stratified and conchiferous tufas, which cover the maritime plains of Western Italy, the Campagna di Roma, and the Terra di Lavoro; and which have penetrated the proximate vallies of the Appenines, then maritime creeks or æstuaries. Pumice, it is well known, floats upon water; when therefore the fragmentary ejections are of this nature, they may be driven by winds or currents often to very great distances, and deposited on the shores towards which the currents set.

If the waters, beneath which the eruption occurs, are impregnated with calcareous matter, the tufas formed in this manner will have a calcareous cement, and will contain sea-shells, &c.; or will occasionally alternate with the limestone deposits of the surrounding ocean.

In this manner were produced the calcareous and conchiferous Peperinos of the Veronese, Vicentino, and the Euganean hills; of Southern Sicily; those of the fresh-water basin of the Limagne D'Auvergne, the Puys de Menton, de la Poix, &c. &c.

On all these points basalt, as well as calcareo-basaltic conglomerate, alternate repeatedly with regular strata of limestone.

When the comminuted and pulverized ejections are of such a nature as to mix into a paste with water, it appears that the mass constituting the cone, and which, from the vicinity of the vent, is intimately mingled by frequent agitation with the surrounding water, and perhaps influenced by the intense heat emanating from the volcanic orifice, assumes a greater consistency and solidity than the strata which are deposited as sediments at a distance from the volcanic mouth; or even on the surface of the cone itself, after the cessation of the volcanic explosions.

Thus the tufa which constitutes the mass, or nucleus, of the eminences surrounding the submarine craters of the Phlegræan fields is sufficiently solid to be worked as a building stone; while the intervening flat spaces consist of loose tufa, identical in composition with the other, and differing only from it in the incoherence of its parts. These beds of loose tufa are scattered also on the surfaces of the harder cones. The difference of solidity between these varieties appears to consist in the former having *set*, or effected a degree of attractive cohesion, during its desiccation; in consequence, probably, of having been intimately mingled into a paste with water by the violent agitation of the fluid surrounding the vent.

The solid tufa, or *trass*, of the Rhine volcanos, is precisely similar in coherence and quality with that of the Phlegræan fields, and owes also its consistency to being violently agitated with water, though under different circumstances, as has been seen in a former chapter. Even some of the finest ashes ejected by the eruption of Vesuvius in 1822, and carried down by torrents over the slopes of the mountain, caked together, on desiccation, into masses of such solidity, as to require a sharp blow of a hammer to break them. They also occasionally assumed a prismatic configuration on the small scale.

Lithoidal Products of Subaqueous Eruptions.

There can be no doubt that the same laws will influence the disposition of lavas produced below any body of water from a subaqueous volcanic orifice, as when emitted in open air. They will in the same manner spread laterally beneath the cover of a scoriform envelope, with a rapidity, and to an extent, proportioned to the expulsive force, their fluidity, and the permanence of that fluidity, and the accidents of level in the surrounding surfaces.

Of these circumstances the first, viz. the expulsive force, depending chiefly on the energy of the eruption, is not probably influenced in any way by the pressure of the upper column of water, since the augmentation this produces in the expansive force is balanced by the corresponding increase in that of repression by which the energy of its development is diminished in an equal ratio. The same may be said of the actual fluidity of the lava, at the moment of its emission from the vent; but the *permanence* of this character is materially affected by the overlying body of water. The great density of the medium with which the liquid lava is brought in contact, by the resistance it offers to the escape of its elastic vehicle, must have the effect of prolonging its fluidity, and consequently, *ceteris paribus*, must augment the extent of its lateral spread upon the same inclination of surface.

Thus we should expect lava beds produced at the bottom of the sea to exhibit a greater lateral extension compared to their thickness, than those which have flowed under the pressure of the atmosphere alone; and that this extension should be proportionate to the depth of the column of water they supported. This reasoning is strongly confirmed by the universal remark of the great horizontal dimensions of those lava beds which are considered as of submarine origin; such as the generality of flœtz-trap formations; e. g. those of Iceland, Ferroe, Ireland, the Hebrides, Germany, &c.

Again; it must be expected that currents of lava which have flowed at great depths under water, will present comparatively few scoriform parts; the surface of the lava being instantaneously consolidated, and the escape of the vapour bubbles, (by the ascending force of their inferior specific gravity) rapidly checked, by the intense external pressure acting on the surface of the current; and this almost total absence of scoriform surfaces has been so generally observed amongst the older submarine traps, as to be a frequent cause of perplexity to geologists.

Vesicles or air-bubbles, on the contrary, should be expected to abound in the interior of the rock, whenever its liquidity was sufficient to permit the agglomeration of the vapour into parcels; the extreme tension of the elastic fluid causing the expansion of the bubbles as the lava flows on; while, for the reason mentioned above, very few will make their escape by rising outwardly.

Hence the numerous vesicular lavas, or amygdaloids of the flœtz-trap formation.

I have already noticed, in a former chapter, the occasional occupation of the vesicular cavities and pores of a lava-rock subsequently to its consolidation, by crystallized minerals, deposited from water percolating through these interstices; and it is obvious, that a bed of lava, existing for a long period under the surface of the sea, and exposed to an intense pressure, will be far more liable to the percolation of water in this manner through its substance, than when consolidated in air. As those elastic fluids, which remain in it subsequently to its consolidation, are slowly condensed by refrigeration, they tend to produce a vacuum in the various cavities they occupy, and oppose no resistance to the penetration of water from above, which is urged to descend, not by its own weight alone, and that of the atmosphere, as in the case of a subaerial rock, but by that of the whole supported column above, in addition.

It is also obvious, that the various mineral substances contained in the water of the sea, by entering into new combinations with the silex and other substances, volatilized, or suspended, in the condensed vapours of the lava, may occasion the crystallization of new minerals within the vesicles and minute interstices of the rock; and this appears to be the origin of the numerous varieties of zeolites, and other minerals, that characterize the amygdaloidal basalts and traps, of which the greater number have been certainly produced by subaqueous volcanos. The slowness with which these substances are separated from their aqueous menstruum, is probably the cause of the great regularity and perfection of their crystalline form.

§. 7. In the case of subaqueous eruptions it will be rarely possible, of course, to distinguish whether they take place from a new or an habitual vent. Both cases, no doubt, will occur, as on terra firma.

In the neighbourhood of Milo, one of the islands of the Archipelago (itself a polystomalous volcanic cone rising from the depth of the sea), there exists certainly a submarine habitual volcano, which, at different epochs, produced the two isles Thera and Therasia (now Santorini), Jera, B.C. 106, Thia, A.D. 4; another near Thera, A.D. 47; in 727 united the two islets of Thia and Tera, and enlarged it in 1427; produced in 1573 the isle called the Little Kammeni, and added to it in 1707. The insular cones thrown up by these different eruptions, are so near to each other, that they must be supposed parasitical or subsidiary apertures of the same fundamental volcano.

On the other hand, the Phlegræan fields, as the volcanic district of Pozzuoli and Cumæ, near Naples, have been called, present an instance of numerous submarine eruptions, each from a fresh point on a shallow shore. Some of the cones left by these phenomena are very regular and entire; others present lengthy circuitous ridges, embracing crateriform basins, of which some contain lakes. They are all formed of an indurated felspathose tufa, containing occasionally sea-shells and fragments of wood, both unmineralized, and are covered by strata of loose tufaceous conglomerate, similar to that which is generally dispersed over nearly the whole neighbouring plain of Campania, and which appears to have been deposited there at a time when the Mediterranean washed the foot of the nearest Apennine.

§. 8. When, at length, the summit of an habitual volcanic cone is permanently elevated above the surface of the sea, it enters into the class of subaerial volcanos, and produces one of those volcanic islands which are of such frequent occurrence in the Atlantic, Pacific, and Indian Oceans.

But it is not solely by the accumulation of matters eructed by its successive eruptions that this increase of height can take place. It may be produced without any addition of bulk to the mountain itself, by its elevation *en masse* from below, occasioned by the general expansion of the subterranean bed of lava, which may also raise, together with the volcanic cone, a greater or less superficial extent of the neighbouring strata forming the pavement of the sea.

This has probably happened to the coast of Italy, just described. The volcanic cones of the Phlegræan fields have certainly sustained a great change of level, relatively to the surface of the sea, since their formation, but without having received any addition to their bulk by subsequent eruptions. Again, the flanks of Ætna are in many parts formed, to a very considerable height, of alternating beds of lava and tertiary limestone full of shells.

We are thus led to distinguish between the islands of volcanic origin, or those parts of them which were produced by subaqueous eruptions, and subsequently elevated above the sea-level (with or without any superincumbent calcareous or arenaceous strata), by successive expansions of the subterranean lava-bed; and those islands, or their parts, which have progressively risen above this level, by the gradual accumula-

M 2

tion of matter expelled during repeated eruptions from an habitually active vent, and in which this process has continued long after their emerging into open air.

§. 9. The latter will have all the characters of a volcanic mountain, viz. a pyramidal or conoidal form, and a structure of alternate beds of lava-rocks and conglomerate, dipping more or less rapidly away on all sides from the central summits.

The former class of volcanic formations will seldom exhibit this regularity of form and structure. In fact, the lava beds which are produced at any depth under the sea level, spread, as we have observed, to a far greater lateral extent than those emitted in open air; and, again, the abrasive action of the marine currents will generally disperse the accompanying fragmentary ejections to a greater distance from the vent. Hence, a submarine volcanic mountain will not present by any means so decidedly conoidal a form as one thrown up in open air, but will be far more depressed and flattened, and composed of comparatively horizontal beds.

Again, the forcible elevation of such a mountain, or of any part of it, by subterranean expansion, will probably still further derange its figure; and, in lieu of a gently sloping conical mountain, as in the prior case, its outline will probably present vast plateaux with little or no inclination, enclosed by precipitous cliffs, or separated from each other by deep fissure-like chasms; and composed of alternate beds of lithoidal and fragmentary lava-rocks, exhibiting the marks of their submarine origin, as well in their great compactness of texture, in the absence of scoriform parts, and perhaps in the numerous amygdaloidal minerals they enclose, as in their peculiar disposition and general forms.

The island of Teneriffe will present us with an example of both these volcanic formations geographically united. Its principal mass consists of an immense volcanic mountain, of which the central cone, or peak, reaches an elevation of 12,200 feet above the sea. In form, this mountain approximates to an oblong cone, although its regularity has been greatly disturbed by the products of uumerous lateral eruptions. The principal crater evidently once occupied the centre of the ellipsoidal base, and corresponded with it in shape and dimensions. Of its enclosing walls, one half only, as has been already said, remains, in a vast semicircular range of cliffs, within the area of which rise the more recent cones of the Peak and Chahorra. But this immense mountain constitutes little more than two-thirds of the superficial extent of the island. On its north-east side, and in the prolongation of the longer axis of the ellipsoid, projects a remarkable headland, differing from the other part totally in form, structure, and composition; having rather a flat and terrassed than a conical outline, and being made up entirely of vast horizontal beds of basalt and basaltic conglomerate, while the rocks of the volcanic mountains are wholly trachytic.*

Fig. 26.

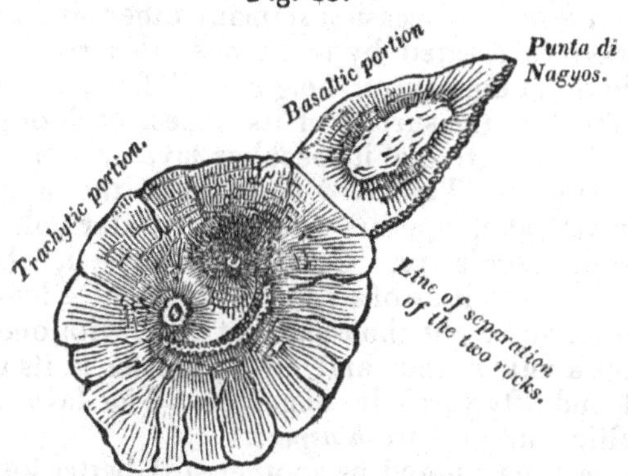

SKETCH OF A MAP OF TENERIFFE.

The basaltic district is clearly a submarine volcanic mass, elevated above the sea level subsequently to its production, since which time the proximate trachytic volcano has alone remained habitually active, and continues so to the present day.

The island of Palma, of which the following cut exhibits a rude sketch, another of the Canaries, offers an excellent example of a perfectly regular and complete insular volcanic mountain.

Fig. 27.

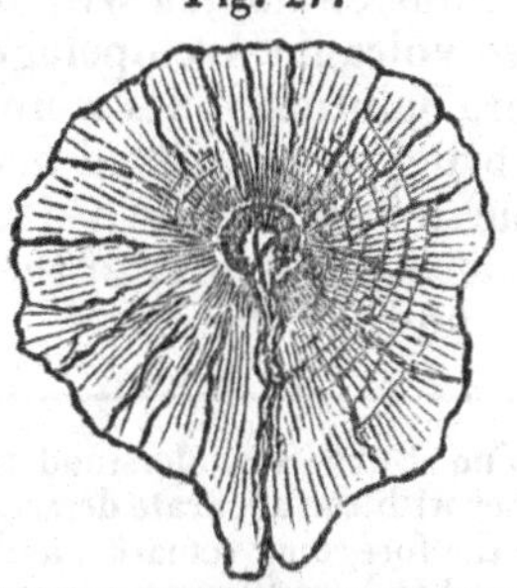

* Von Buch ueberden Pic von Teneriffe.

Its form is perhaps as near an approximation to the cone as the disturbing causes which must necessarily influence the figure of such a mass will ever permit; the outline of its ground plan is nearly circular, and it rises from the coast on all sides, at first by gently inclined slopes, which become gradually more and more rapid, till they terminate in the ridge which forms the summit of a steep range of cliffs encircling its central crater. This deep cavity is drained by one outlet, perhaps originally a fissure, such as were described on a former occasion in many other similar localities. The ravines created by numerous other rivulets, diverge like radii from the central heights. A few parasitical cones are sprinkled on the surface of its slopes, each of great regularity, and productive of its peculiar lava current.*

The island of Tristan d'Acunha, in the middle of the Southern Atlantic, appears to be an insular volcanic mountain of equal regularity. Captain Carmichael, who visited it in 1816, describes it as of a conical figure, nine leagues in circumference, and eight thousand feet high, with one large central crater a mile round, and having a lake at its bottom one hundred and fifty yards in diameter. The lava and scoriæ are basaltic, and of a *fresh aspect.*†

Ischia may be quoted as another still better known example. Its form is sufficiently regular, though disturbed by numerous parasitical cones, and the lithoidal products of lateral eruptions. At its summit are evident traces of a great central crater, of vast size, which appears to have been long extinct. The lavas produced by this volcano, have oscillated between trachyte and basalt.

In fact, such appears to be the general form of all those insular volcanos, which have been chiefly produced by habitual eruptions from a central vent, since its elevation above the sea-level; and numerous examples will be found, no doubt, amongst the immense volcanic Archipelagos of the Pacific.

Where two or more habitual vents were situated so near to one another, as to bring their erupted matters into contact, the island will of course be formed of a string of such conoidal mountains; and numerous instances of this are to be met with in the Leeward Isles and the Pacific.

---

* The maps which Von Buch is understood to intend publishing of these two islands, together with the accurate descriptions of that celebrated geologist, will illustrate the foregoing remarks, and supply the deficiencies of the brief description, and rude outline plans given above.

† Trans. Linnean Soc. vol. xii. part 11.

Even Java, a great part of Sumatra, Celebes, Japan, the promontory of Kamskatcha, and many of the larger islands of this part of the globe, appear to consist of one or more rows of volcanic mountains thrown up from vents produced upon the same or parallel fissures in the crust of the globe.

The eruptions of such habitual volcanos, will, no doubt, have been often accompanied by contemporaneous local elevations of the neighbouring strata by the force of subterranean expansion; and hence strata of limestone, sandstone, and even of granite or the schistose rocks, may form a part of these islands without controverting in any way the origin here attributed to them.

As examples of islands, composed chiefly of volcanic products, which appear to have risen from below the sea, solely by subterranean expansion, without having been since augmented in height or bulk by the products of external eruptions, we may perhaps venture to name the Madeira and Ferroe Islands, some or all of the Trap Islands amongst the Hebrides, those of Ponza, Zannone, and Palmarola, on the coast of Italy, &c.

Iceland, like Teneriffe, presents a union of both kinds. The high mountains called *Jokuls*, with which this island abounds, marking the site of so many habitual volcanos, which continue still, or have lately been, in occasional activity; while a large part of the island consists of flat plateaux, composed of repeated beds of basalt and basaltic conglomerate (basalt-tuff and trap-tuff—Mackenzie), bearing many marks of a submarine origin, and of their having been subsequently raised to their present level.

The isles of France and Bourbon present a similar contrast, the former having all the distinguishing characteristics of a submarine volcanic formation, elevated since the cessation of the eruptive phenomena; the latter, of an ordinary volcanic mountain, produced by repeated extravasations of lava from two or three habitual sources above the sea-level. One of these vents has remained in continual activity to the present day.

Where a volcanic mountain has been elevated, *en masse*, by inferior expansion, it will be of course accompanied in its rise by the recent beds of marine formation which had been deposited upon it; and even strata of earlier date may occasionally be raised with it by the same expansive force acting upon a more considerable extent of surface, than that which is loaded by the volcanic mountain.

Thus, the northern extremity of the Isle de France presents a flat plain, composed of a calcareous rock full of recent madreporites, and other coralline bodies, overlapping the volcanic rocks which make up the remainder of the island. The neighbouring islets of La Platte and Les Colombiers offer a similar conformation.

So, again, in the cluster of islands lying eastward of Java, the greater part of their surface is composed of coralline beds, unmineralized, and exactly similar to those which constitute the neighbouring reefs still in progress of formation below the level of the sea. These beds visibly rest upon volcanic rocks, which have in this instance been, no doubt, elevated, *en masse*, together with the coral reefs which had been built upon their highest eminences.

The island of Pulo Nias, on the western coast of Sumatra, which is seventy miles long and twenty-five in width, exhibits the same fact throughout its whole extent, with this exception, that the coral beds rest on stratified sandstones and limestones of older character. The nearest coast of Sumatra is, however, volcanic; and the elevation of these extensive recent coral beds, to a height of some hundred feet, is most probably owing to subterranean expansions coetaneous with the volcanic phenomena of the neighbouring island.*

There seems, indeed, great reason to believe the almost innumerable coral reefs of the Pacific and Indian Oceans to be generally based upon the summits of submarine volcanic mountains.

Their figure is usually circular or elliptical, corresponding to what we may suppose the ridge of the central crater. By the ordinary process of growth, so well described by Dr. Macculloch (Journal of Science, January 1823), their mass cannot attain, at the utmost, to any greater elevation than a few feet above high water mark; whereas, in the numerous Archipelagos of these vast oceans, a very great number of islands, which are shown by their composition and structure to have been originally formed as coral reefs, rise much above this level, attaining frequently from two to three hundred feet in height, and occasionally much more.

Those, however, which have reached this height, have been observed generally to consist of a substratum of basaltic lavas, supporting the coralline beds; and it is a common remark of navigators, that they are subject to frequent and violent earthquakes. All these facts combined, make it difficult to doubt, that these islands are the summits of subaqueous volcanic cones, which, when raised to within a certain distance of the

* Dr. Jack, on the Geology of Sumatra.—Geol. Trans. (second series), vol. i. part II.

surface of the ocean, are immediately occupied by the remarkable zoophytes which elaborate coral, for the site of their erections; and this distance is not considerable, perhaps never exceeding a hundred feet,* these animalculæ requiring light, as it would appear, for their support.

Subsequently, by subterranean intumescence, acting probably by repeated shocks, the mountain is more or less elevated, and by the continual growth of fresh coral on its shelving shores, is progressively augmented to a considerable size.

If we look at the prodigious number of coral reefs and islands that stud the Pacific, and reflect that the process above described, consisting in the creation of new islands, and the elevation and augmentation of those already existing, is unceasingly carried on, we shall be led to suspect that we are perhaps witnessing the operations of nature in the slow elaboration of a new continent, which may one day rise in the place of the Polynesian Archipelago.

It appears from the accounts given by M. Moreau de Jonnès and others, of the structure of the Leeward Isles, that they offer another striking example of this association of elevated and erupted rocks. Those islands which lie farthest to the west, or "under the wind," are described as consisting solely of volcanic cones, strung together at greater or less distances; while the eastern line of islands is formed of calcareous strata of different ages, and finally of recent coral, supported on a foundation of trachyte, and other volcanic rocks, which may therefore fairly be presumed to have been elevated, together with the overlying strata, by subterranean expansion.

We shall revert to these proofs of the elevation of marine formations, and other non-volcanic rocks, by subterranean expansion, at a future opportunity.

---

* See an excellent memoir on this subject, read by M. Quoy to the French Institute, July, 1823. M. Quoy, from the result of his observations (on board the corvette Uranie) asserts the extreme depth at which the Madrepores subsist to be thirty feet.

CHAP. IX.

Systems of Volcanos.

§. 1. Whenever a new vent is broken through the superficial strata of the globe, in consequence either of the efficient obstruction of some of the neighbouring apertures, or their insufficiency to discharge all the redundant caloric communicated from below to the subterranean bed of lava; the situation of this fresh opening relatively to the other prior vents, will be determined by the local circumstances of tenacity, and solidity, in the strata against which the force acts;—the fracture taking place wherever these strata are weakest.

If by the expansive efforts which preceded or accompanied the eruptions of the more ancient vent or vents, the overlying strata should have been, as is probable, so much fractured and displaced, as to render them weaker and less efficiently resistant in the neighbourhood of these points, than on their more distant and intact parts, the fissure of eruption will be produced somewhere in the immediate vicinity of the older volcanos.

The general prevalence of this law has, without doubt, given rise to the numerous groups or systems of volcanic vents which are observable on the globe. Of these, some are united into clusters or detached irregular groups, the position of each vent bearing no very apparent relation to that of the others; such are those of Iceland, the Azores, Canaries, and Cape Verd Isles, in the Atlantic; and the Archipelagos of the Moluccas, and Gallipagos, in the Pacific, &c.

The generality of groups, however, have a decided *linear* arrangement; one vent following the other in the continuation of the same straight or nearly straight line; and when volcanos have been formed on neighbouring points out of this principal line, they are in almost all cases situated upon other rectilinear bands parallel to the first.

The cause of this remarkable conformity cannot be mistaken. It has been observed above, that every fresh vent or fissure of eruption must be produced on that point where the strata overlying the subterranean lava mass yield most readily to its expansive efforts; and there can be no doubt that, in the generality of cases, this point will lie *on the prolongation of the original fissure* from which the older volcanos were thrown up. But this fissure, when first formed, took the direction in which these strata split most readily; consequently whatever other clefts are occasionally formed through the same strata, but out of the line of direction of the former fissure, will be most probably parallel to it; the direction of the later fractures being influenced by the same circumstances which determined that of the former.

The operation of this law is so general that it may be almost said that wherever volcanic vents have been produced, there we may perceive an example of this linear arrangement. Some of the most remarkable series of this kind may be briefly noticed.

Frequent mention has already been made of the two parallel ranges of extinct volcanos, most of them productive of but a single eruption, which stretch across the middle of France in a direction nearly north and south, and to a longitudinal extent of more than two degrees. Three colossal volcanic mountains, the Monts Dor, Cantal, and Mezen, the two first being sufficiently regular in figure and structure, mark the site of as many vents of habitual and long continued eruptions on different points of these zones.

Another linear series of extinct volcanic vents runs across the north of Germany from west to east, commencing in the cluster of the Eiffel, and terminating in the Mittelgebrige; unless it is allowable to suppose it continued in the volcanic groups of Hungary beyond the Sudetengebirge, and Carpathians. The volcanic districts of Italy are stretched upon a nearly continuous line from the Montamiata on the north, through the Montecimini, the Campagna di Roma, and the Valley of Pofi and Frosinone, to the Phlegræan fields of Naples, which, by means of the Lipari Isles, are connected with Ætna and the Val di Noto. Another range runs parallel to this, for some distance, in the Isles of Ponza, Ventotiene, and Ischia.

The Antilles or Leeward Isles offer a similar train of volcanic openings, most of which are extinct, although some occasionally give rise to rare and inconsiderable eruptions. According to Moreau de Jonnès, the islands of Saba, St.

Eustache, St. Christopher's, Nieves, Montserrat, Guadaloupe, Dominica, Martinico, St. Lucia, St. Vincent, the Grenadines, Grenada, and Trinidad, are wholly volcanic, forming a chain of vents 200 leagues in length, and running in a direction nearly N. and S. Some of these islands consist of two, three, four, and even more volcanic mountains, ranged on the same meridian, and connected at their bases. Martinico has six of these conoidal eminences, the sites of habitual volcanoes, which have been long quiescent. Nearly all have central craters sufficiently distinct.

A still more striking and constant allineation of volcanic spiracles is displayed in the numerous Archipelagos which border the eastern coast of Asia.

Beginning from the Peninsula of Kamstchatka, itself a range of similar mountains, the islands of the Kurille group, Jesso, Japan, the Loo-choo, Phillipines, and Moluccas, form a scarcely interrupted linear series of volcanic islands; amongst which many contain vents at present in habitual activity, while all exhibit a greater or less number of conoidal volcanic mountains, arranged sometimes on one, sometimes on parallel zones, and bearing marks in general of recent and extremely violent eruptions.

But this is not all; for after expanding to embrace the whole group of the Moluccas, this chain of vents branches off in two directions towards the east and west. It is continued in Timor, Sumba, Sumbawa, Java, Sumatra, and probably the Nicobar, and Andaman isles, to the western coast of the Burmese empire, on the one side; on the other, crosses New Guinea, and stretches into the centre of the Pacific, giving rise to the Archipelagos of New Britain, Solomon, Holy Ghost, Friendly, Society, and Dangerous Isles, &c. &c.

A glance at the direction of these linear groups on the globe will be sufficient to disprove the strange assertion occasionally put forth by Moreau de Jonnès and others, of their being universally coincident with the Meridians.

The chain of the Aleutian isles in particular, which is entirely composed of volcanic mountains closely linked together, is, on the contrary, directed precisely from east to west.

The last mentioned range of volcanic vents, which on the west connects itself with the Kamstchatkan volcanos, is continued at its eastern extremity in the neighbouring promontory of America, from whence there is great reason to suppose that a more intimate acquaintance than we possess at present with the north-west coast of this continent, will shew it to be continued along the shore of the Pacific into California, where many habitual volcanoes are known to be situated, and thence through the heart of Mexico, Guatimala, Nicaragua, and the isthmus of Panama, to the vast volcanic formation of the Andes of South America, which reaches uninterruptedly to Terra del Fuego, the southern extremity of this remarkable continent.

So extensive a train of volcanic vents, stretching in a curved line across the surface of the globe, from 10° S. latitude to 60° N.; and thence returning again to nearly 60° of south latitude, is a fact too remarkable to be overlooked, or only cursorily glanced at. The cause which determined the formation of so prolonged a fracture through the crust of the globe in this direction, is beyond our reach, but the results of this circumstance may perhaps be perceivable in some of the characteristic features of this hemisphere.

Indeed so generally does this linear arrangement prevail in the disposition of volcanic vents on the surface of the globe, that even detached single volcanos or clusters of volcanos, appear frequently connected as it were by intervening points on which volcanic vents have once existed, or may be suspected to exist, beneath the surface of the ocean.

Thus the Madeiras form an obvious connecting link between the clusters of the Azores and Canaries; and suggest the idea that other links of the same chain may exist unknown to us below the level of the sea.

The same may be said of the Ferroe Islands, with respect to the isolated groups of Iceland on the one hand, and the Hebrides and north of Ireland on the other.

§. 2. In these systems, whether linear or grouped, one or more different vents remain perhaps in occasional activity, while the others in which the repressive forces have permanently acquired the predominance, appear extinct.

Thus the active volcanos of a system may be reckoned as safety valves, which continually, or by successive eruptions, letting off the excess of caloric communicated from the centre of the globe to the subterranean bed of lava, proportionately obviate its expansion, and prevent the occurrence of those forcible ruptures and elevations of the overlying strata, which occasion destructive earthquakes, and alter the relative level of land and water.

Should these spiracles become gradually obstructed by their accumulated products, or other causes, such heavings and ruptures must recommence, and continue at intervals with greater or less violence, until either entirely new orifices are opened in the vicinity, or those which have long lain dormant are re-opened.

It will be however impossible in any individual instance to calculate the epoch of such an occurrence, so many of the circumstances which make up the sum of the force of resistance being beyond the reach of our investigations.

Many facts may be brought forward in illustration of this reasoning. Thus, for example, the train of volcanic vents of the Leeward Isles, which bear signs of having existed in full and vivid activity for a long period, are now become so sluggish as to offer but two or three eruptions from any volcano in the course of a century, and these, when they occur, are comparatively trifling and powerless.

But, at the same time, this chain of islands, as well as the neighbouring coast of Cumana, the Caraccas, and Venezuela, has long been subject to frequent and violent earthquakes, which are observed to produce numerous rents or fissures through the solid rocks composing these districts, *and to elevate them perceptibly at each successive shock, above their former level.**

These two general circumstances, viz. the rarity and insignificance of the eruptions from these spiracles, and the frequency and violence of the most terrible class of earthquakes, are obviously connected in the relation of cause and effect. Thus again the Andes of the province of Lima, in which there is at present but one volcano in activity, and that by no means energetic, while the neighbouring districts of Quito to the north, and Chili to the south, present several, are grievously disturbed by earthquakes, which often occasion deep and extensive clefts through the superficial rocks, so wide that they can be crossed only by bridges.†

It is a remarkable fact, that the inhabitants of Mindanao, one of the Moluccas, are so convinced by experience that the habitual activity of the volcanic vents in their island obviates the destructive earthquakes which would otherwise ruin their towns, that whenever these volcanos are quiescent for a longer

* See Mrs. Graham's account read before the Geological Society. The connection of these earthquakes with the volcanic vents of the Leeward Isles was further proved by the eruption of St. Vincent's in April, 1812, but a few days after the terrific earthquake which desolated the provinces of the Caraccas in the same year.

† Mr. Caldcleugh says expressly, that of late years while there has been an increase in the number and violence of earthquakes, the eruptions from the volcanic chimneys of the Andes, have been proportionately rarer.

Travels in S. America, vol. ii. p. 48.

period than usual, they sacrifice a slave to appease what they suppose the wrath of the Deity that inhabits them. The same opinion is reported to prevail among the natives of Sumatra.

§ 3. The continuance of the process of elevation, thus occasioned by the absence or obstruction of eruptive vents which might relieve the subterranean lava bed of its redundant caloric, when operating on any extensive mass of the overlying strata for a great period of time, must raise them to a proportionate height above their former level, and at the same time give rise to numerous fractures, dislocations, and irregularities of position.

We have no reason to doubt that the transmission of caloric from the central parts of the globe towards its circumference, to which we have been led to refer the phenomena of volcanos and earthquakes, has been constantly in operation, at least with equal, if not greater energy, since the infancy of our planet.

Consequently we should expect to find in the constitution of the crust of the globe indications of such changes of level in the superficial strata composing this crust, and frequent traces of the fractures, dislocations, and irregularities of position which must have been occasioned whenever they were elevated in an extreme degree.

How completely the appearances presented by the strata of our continents, and more particularly those which compose their most elevated or mountainous districts, correspond with what would thus appear à priori to be the necessary results of the process in question, must be obvious to every one.

The inclined position of these strata, their manifold inflexions, the mutual intersections of their fractures, the veins, shifts, or faults, that interrupt their continuity; all which marks of disturbance encrease in frequency and evidence as we approach the mountains, that is, *those points which, under our hypothesis, have sustained the maximum of elevation*; while, on the other hand, the composition and contents of these strata prove that the greater part of them originally existed at the bottom of the sea in a horizontal and continuous position; these and the numerous other circumstances which it is needless to repeat, since they have been so ably and eloquently enumerated by the advocates of the Huttonian theory, compose an irrefragable mass of evidence in favour of the progressive elevation of our continents by a force acting from below upwards.

Here then we have a series of effects completely adapted and proportioned to that cause, which the phenomena of volcanos prove to be in constant operation.

On the one hand vast masses of the globe's crust exhibit incontestible signs of their having been gradually raised by the successive efforts of a powerful force acting from below them.

On the other the still active volcanos scattered over the earth, and the frequent traces of their former activity, testify to the permanent existence of a subterranean expansive force to which we can perceive no limit, and which, on those points where the caloric, by which it is animated, is prevented from escaping through these accidental spiracles, must necessarily give rise to successive violent elevations of the superficial strata on the most extended scale.

It would be wholly unphilosophical to withhold our belief that these circumstances are connected in the relation of cause and effect. We must conclude then, that the successive local expansions of the general subterranean lava bed have occasioned the present height of the continental rock masses above the level of those which underlie the ocean. The eruptions of volcanos are only a secondary phenomenon, the partial and incidental result of these elevations.

§ 4. But since we have no reason to doubt the influence of subterranean expansion in the phenomena of volcanos, earthquakes, and elevations of the superficial strata, to have taken place for ages past, indeed to have been co-eval with the existence of our planet, under its actual laws, if it be true that the development of the one class of these phenomena, viz. volcanic eruptions, proportionately obviates that of the other, or the absolute elevation, *en masse*, of extensive superficial portions of the earth's crust; and therefore in the same locality, and at the same period, the one class of effects must always have varied inversely with the other; we should expect to find proofs of the operation of this law in the visible traces left by these phenomena on the globe, and, consequently, that, wherever volcanos have existed in the greatest number, and in the most prolonged and energetic activity, there will have happened the least absolute elevation, *en masse*, of the superficial strata; and, vice versâ, that wherever the most remarkable elevations of the solid crust of the globe have taken place, there should seem to have been little or no absolute escape of subterranean caloric through the spiracles of habitual volcanos.

What are those spots, on the surface of the globe, where the volcanic action appears to have been most general, most frequent, most energetic, and most prolonged? Are they not all either maritime coasts, or islands rising from the depths of the ocean? Look at Iceland, where more than twenty habitual volcanic vents appear to have been almost incessantly in active operation, and where there are scarcely any signs of elevation.

None of the pre-existing marine strata appear in this case to have been raised above the sea-level. No primitive, transition, or secondary rocks are, I believe, met with in the whole extent of the island.

The same remark is applicable to the Azores, Canaries, and Cape Verd Isles. But above all, let us turn our attention to the numberless volcanic vents that stud the Pacific.

At the first glance on this quarter of the globe it is impossible not to be struck by the total absence of any continent whatever, and the extreme scarcity of high and dry land, (for most of its islands are mere coral reefs) throughout the vast extent of ocean, which alone occupies a full third of the superficies of the habitable globe. At the same time we observe this immense basin to be almost encircled by an extraordinary chain of closely grouped volcanic vents, described above; of which very many are still, and a much larger number have certainly once existed, in habitual, prolonged, and exceedingly energetic activity; while, the interior of this ocean is thickly sprinkled with groups of similar spiracles, (the Marianne, Carolinas, Sandwich Isles, &c.); and the frequent production of new islands and coral reefs, warrants a belief in the existence of numerous others which have not yet showed their summits above the water-level.

It is in this quarter of the globe, without doubt, that the volcanic phenomena have been most copiously developed; and it is in the same quarter that so vast an extent of the surface of the earth appears to have preserved almost entirely its original level, comparatively uninfluenced by that elevating force, which on other parts has occasioned the production of the colossal continents of the old and new world.

This remarkable correspondence speaks loudly for the correctness of our argument.

But it may perhaps be urged that of the great linear chain of vents described as almost encircling the Pacific, one very considerable portion belongs to the elevated continent of America; and that, whatever argument may be drawn from the absence of any elevated extent of the superficial strata on the *one* side of these volcanic trains, must be opposed by their occurrence on the *other*.

To this objection, which is striking and plausible enough at first sight, the following considerations will be a sufficient answer.

In the first place, the stupendous mountain masses of the Andes and Cordilleras of Mexico and South America, upon which the strength of the objection mainly rests, are, it must be recollected, themselves of volcanic formation, consisting of immense bodies of lava protruded from the numerous and extremely productive vents which have been opened along this extensive range; and the few portions of elevated strata which are met with in their vicinity, rarely attain to any great height, and form but a very inconsiderable proportion of these enormous mountains.

In the second place, we may remark that no volcanic vent, or series of vents, *can be produced*, but upon a fissure or solution of continuity, effected in the superficial strata of the globe, *during their elevation* by the force of *subterranean expansion.* Consequently, a certain degree of elevation of the neighbouring strata, *on one side or other* of the fissure, *must precede the establishment of any* train of volcanic spiracles, and also *accompany the further development* of their phenomena. And with respect to any single fissure or cleft of this nature, the greater the elevation produced in the neighbouring strata, the wider and deeper will be the fissure, and the greater the space afforded for the escape of volcanic matter.

In the case of America, for example, if we suppose the great compound fracture which produced the linear series of vents, already described as reaching perhaps from Icy Cape to Terra del Fuego, to have been originally occasioned by the incipient elevation of the continents of North and South America; and to have been subsequently widened and prolonged, as this extensive mass of indurated strata rose progressively above the sea, and acquired its present level, we shall see at once the cause of the enormous quantity of lavas which have been emitted successively from this cleft; and the existence of these elevated continents on the *one* side of this linear fracture, will not be found in any way to contradict the opinion that the activity of its volcanos has mainly contributed, by letting off the superfluous caloric from the bed of lava situated beneath the Pacific, to obviate the elevation of its bottom on the *other* side, and thus occasion the remarkable scarcity of land to the west of the Andes and Cordilleras.

The accompanying diagram, presenting a rude ideal section of South America, will illustrate the position better than any verbal argument.

Fig. 28.

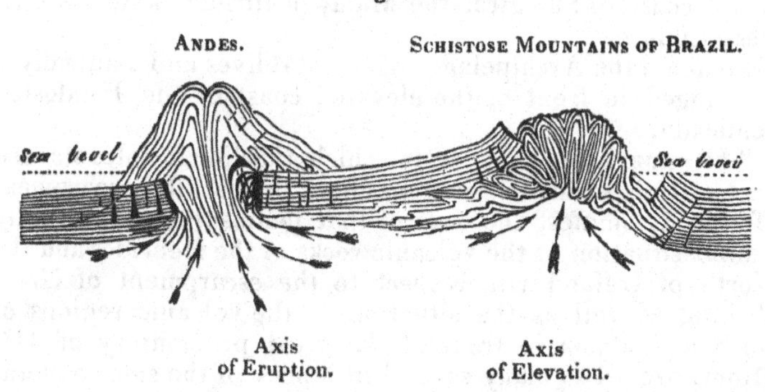

The remarkable absence of all volcanic vents, and, I believe, even of all traces of their existence at any former period, on the Eastern side of the Andes, and their prolongation, the rocky mountains both in North and South America, affords the strongest confirmation of the truth of the above conjecture, as well as of the genuineness of the law which we are investigating.

If we turn to the old world we observe a repetition of these circumstances. In fact, the linear chain of volcanic vents which borders the Pacific on the East, follows in a most remarkable manner the outline of the Asiatic Continent, and leads to the supposition that the formation and enlargement of this lengthened series of fractures in the crust of the globe, accompanied the elevation of that extensive portion of it which constitutes the neighbouring continent.

If from Asia we direct our attention to Africa, it is not difficult to discern in the volcanic islands or insular groups of the Canaries, Cape Verd Isles, Ascension, St. Helena, Saxemberg, Tristan d'Acunha, and Diego d'Alvarez, on the west, and of the Mauritius, Seychelles, &c. on the east, the highest summits of a train of submarine volcanic spiracles, the direction of which corresponds with that of the outline of the neighbouring continent, in the same remarkable manner as in the instances of Asia and America noticed above.

Even many, or most, of the smaller and more partial lines on which the volcanic force has been developed, appear to bear the same striking relation of direction to some contiguous masses of elevated strata.

Thus the volcanic range of Java, Sumatra, and the Andaman Isles, forms a sort of advanced breast-work to the parallel range of primitive and secondary mountains of Borneo, the North coast of Sumatra, the Malay peninsula, and the Burmese Empire.

So again the Archipelagos of the Maldives and Laquedives are ranged in front of the elevated coast of the Hindostan peninsula.

The remarkable manner in which the linear volcanic district of Italy seems to rise from under the parallel escarpment of the Apennines, the axis of that peninsula;* the exactly similar situation of the volcanic rocks of the Hebrides and the North of Ireland with respect to the escarpment of Great Britain, as well as the situation of the volcanic regions of Ionia and Mysia in front of the great promontory of Asia Minor, are all so many several instances of the same general fact, which can be accounted for only by allowing the same cause to have operated both the elevation of the strata, and the production of the fissures of eruption.

This fact is still further confirmed by the observation, which I believe will be found generally correct, that the direction of any range of volcanic vents is parallel to the general direction of the elevated strata in its vicinity. The volcanic zone which crosses the South of France runs nearly north and south, and this is the general direction of the beds of granite, gneiss, and mica schist, which form the elevated plateau from which the eruptions have broke forth.

The volcanized band which traverses the north of Germany is remarkably parallel to the axis of the Alps, and that of Hungary to the direction of the Carpathians.

The basaltic zone of the Hebrides takes the direction general to the strata of the neighbouring gneiss islands; the volcanic trains of Italy, that of the Apennines.†

§. 5. But if it be true that the development of the subterra-

---

* At the time of the earthquakes in Calabria and Messina, the ground was split open by crevices, which were all directed *parallel to the line of the coast*; that is, to the axis of the range of mountains.

† One exception to this general rule is reported to exist in Mexico, where four or five volcanic vents lie on a line running directly across the axis of the mountain chain. These then were formed on a *transverse* fissure.

nean force of expansion in the elevation of strata, has generally varied inversely with its other modification, which consists in the absolute escape of intumescent matter on the surface of the globe, we must also expect on the other hand to find few or no indications of the latter class of phenomena (volcanos) where the superficial rocks have been considerably raised.

And, indeed, the rare occurrence of volcanic vents in the interior of any continent is a circumstance so generally acknowledged as to have given rise to many specious theories to account for it.

We know of no active volcano in the interior of Europe, or Africa, and though two are reported to exist in the centre of Asia, the fact appears by no means certain, and their activity, at all events, reduces itself to the emission of sulphureous and ammoniacal vapours.

We are equally ignorant of the occurrence of any rocks of a volcanic nature testifying to the former existence of such spiracles in the interior of either of the two larger continents of the old world; and though the volcanic remains of France and Germany are placed at a certain distance from the sea, and even have been produced on points of considerable elevation, these examples are to be considered rather in the light of exceptions, and do not militate against the general fact that such formations are comparatively rare in inland positions.

With regard to the western hemisphere, we have already noticed the seemingly total absence of volcanic spiracles, either open or extinct, in North or South America, westward of the Andes, and the continuation of this range in the Mexican Cordilleras and rocky mountains. But, on the other hand, it is a very remarkable circumstance, that exactly where these continents taper off in breadth, and are eaten into by the great Gulph of Mexico, another range of volcanic vents makes it appearance in the Leeward Isles, running nearly in a parallel direction to the very active volcanic train of Guatimala, Nicaragua, and the Isthmus of Panama; this extreme development of volcanic action obviously accounting for the inferior extent of land which has been elevated on that paralle of latitude.

§. 6. But it may be asked how it happens that earthquakes are not found to take place more frequently and with greater violence in the interior of continents, and at a distance from all active volcanos, whereas it is on the contrary a known fact that many of the districts which are most subject to these destructive phenomena, are placed, though not in their immediate vicinity, yet at no great distance from habitual volcanos, e. g. Calabria, situated between Vesuvius, Stromboli, and Ætna; and the coast of Venezuela and Caraccas, opposite to the Leeward Isles, &c.

To this it may be answered,

I. That earthquakes and volcanic eruptions are by our theory jointly occasioned by the same circumstance, viz. the rupture of the solid overlying strata by subterranean expansion. The production of a fissure, and therefore an earthquake of more or less violence must precede every eruption; consequently we must always expect earthquakes to occur in the vicinity of volcanos. But the reverse is not equally true, since any number of crevices may be formed in the overlying strata, without giving rise to a single volcanic eruption, where the focus of expansion lies at a great depth beneath the refrigerated strata; and in accordance with this remark, many places are known to be liable to most destructive earthquakes which are yet far removed from any habitual volcano; for instance, Lisbon, the Ionian isles; and again Syria and Persia, which of late years have suffered in a remarkable degree from these calamities.

II. With regard to the most elevated portions of our continents, the vast expansion which has already taken effect on these spots in the subterranean lava bed has in turn given the predominance to the force of repression, in obedience to the general law laid down in a former chapter, and should by no means lead us to expect the continuance of the process, and therefore of the earthquakes which are its sensible results, but, on the contrary, to suppose that this tendency has already satisfied itself by the intumescence and consequent refrigeration of so vast a body of crystalline rock, the diminution of temperature being always proportionate to the expansion; and also that owing to the greater comparative distance of the surface of the earth on these points from the central focus or source of heat, the caloric can no longer be concentrated to any intense degree beneath this surface, but is drawn off laterally towards those lines of volcanic activity, which, however distant, are yet still occasioning a continual draught of caloric from every side.

A great quantity, it is also probable, escapes from mountainous districts by gradual transmission, or is carried off by the waters which filter through the numerous fractures and crevices of their constituent rocks, and after acquiring the temperature of these surfaces, find their way to the interior of the globe at a lower level.

§. 7. The generalization of this important fact, viz. that the elevation, *en masse*, of the solid strata composing the crust of the earth, has been always inversely proportionate to the development of the *volcanic* phenomena in the same quarter of the globe, demonstrates that the subterranean bed of intensely-heated crystalline rock, whose *local* existence we fully proved in the early part of this essay, but without defining its limits, and from whose gradual increase of temperature, and consequent expansion, these phenomena, of either kind, have necessarily resulted, must extend generally beneath the surface of the whole globe.

The successive expansive shocks of this subterranean bed, by which the continental parts of the earth have been progressively elevated, incidentally gave rise to contemporaneous eruptions, i. e. to partial outward intumescences of this matter, by occasioning cracks or fissures, on the prolongation of the elevated strata, and more or less distant from the line of maximum elevation, of sufficient depth and width to permit of this extravasation, according to the laws investigated in the previous chapters.

The volcanic phenomena are then only *secondary* and attendant circumstances on the more immediate and *primary* results of subterranean expansion, viz. partial elevations of the solid crust of the globe.

The nature of the more important and extraordinary class of effects, we go on to consider in the next chapter.

CHAP. X.

Development of Subterranean Expansion in the Elevation of Strata, and Production of Continents above the Surface of the Ocean.

---

§. 1. The masses heaved upwards by this immediate action of the force of subterranean expansion have been hitherto spoken of under the general term of overlying or superficial strata: but on examining their constitution, we find them to comprehend a variety of different rocks.

The uppermost, on those parts of the continents which have suffered least disturbance or elevation, generally consist of more or less horizontal strata of sands and marles, or sandstones and limestones. As we approach the chains of mountains, or lines of the maximum of elevation and disturbance, these arenaceous and sedimental strata are found to assume a great degree of inclination and numerous other irregularities of position.

They however universally lean against and are supported by masses of *crystalline* rocks which form the geological axis of every mountain chain.

Of these some are stratified, or rather possess a laminated structure; others not. The former, or stratified crystalline rocks, gneiss, mica slate, &c. exhibit great disorder and every mark of forcible elevation, in their highly inclined and generally vertical position, their flexures, solutions of continuity, &c.

The latter, or *unstratified* crystalline rocks, (granite, syenite, porphyry, serpentine, diallage rock, and greenstone, &c.) usually underlie and support the others, or cut through them in the manner of immense dykes.

In fact they have every appearance of being portions of the subterranean crystalline bed, protruded through the stratified rocks, sometimes in a state of partial liquefaction, at others as a solid mass, preserved from ebullition by the immense pressure of the supported strata, or of the ocean above them; a circumstance already mentioned as one of the probable modes of action of the subterranean expansive force.

§. 2. If it is true that these masses were in this manner forced upwards through vast fissures in the overlying strata, which were rent asunder by their protrusion, the fractured extremities of these strata should be found to correspond on either side of the crystalline axis; and this is in fact frequently the case, as in the instance of the Hartz mountain, whose structure was taken by Werner as the type of mountain ranges; but in general great irregularities will be necessarily produced by the violence of the process, and by numerous circumstances connected with the nature of the elevated strata.

There is great reason to conclude that in most instances the raised strata, particularly those which were only partially indurated, have become contorted, and bent into repeated foldings, so as to give the appearance of frequent alternations of different series of strata to what is in reality but the replication of the same original series.

Fig. 29.

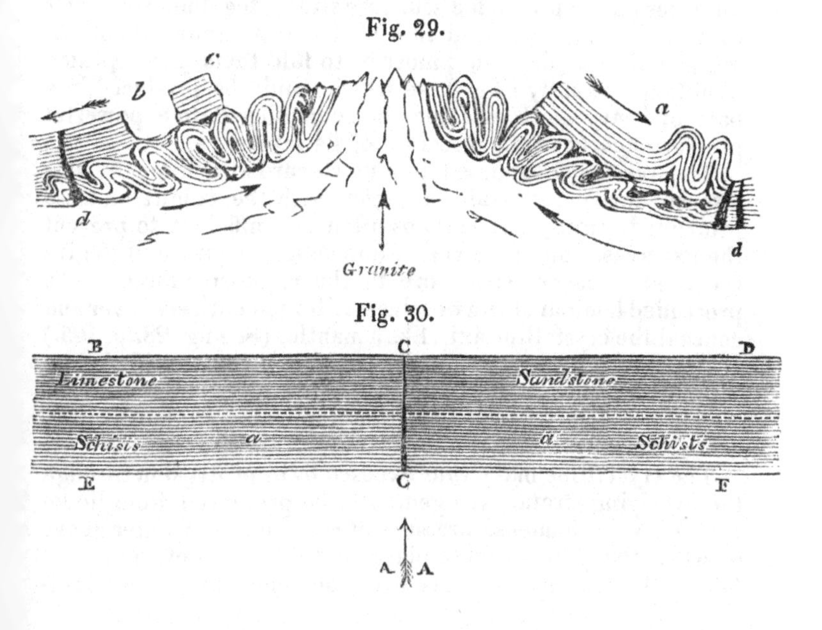

Fig. 30.

If we suppose a fissure C to be broken by the subterranean expansive force through the overlying strata, which are not throughout completely indurated, and therefore possess a certain degree of mobility amongst their parts, the expansion of the crystalline mass A bursts open the fissure C, and elevates the strata on either side with immense violence.

During this process the lower strata E F, which are in contact with the expanding crystalline mass, must sustain an intense friction from it, as it forces its way towards the opening. If these strata are schistose or lamellar, that is, composed of laminar particles, and not entirely consolidated, they have a much greater degree of mobility in the direction of their lamina than in any other, their parallel plane surfaces slipping readily over one another; and this direction coincides with that of the impulse communicated by the friction, namely, from E and F towards C.—Consequently these strata will be forced to move towards the fissure C, and will be protruded from them, together with or even before, the crystalline mass which urges them forward.

But the matter thus forced from opposite sides of the fissure towards the axis, must meet there, and create an intense mutual pressure in a direction nearly opposite to that of the protruding force; and the effect of these two nearly opposite impulses on the protruded schistose strata, together with their own gravitating force and that of the strata above them acting in a vertical direction, must be to fold them into repeated doublings, exactly like those which would be produced in a bale of heavy cloth or linen, by subjecting it to a powerful pressure nearly in the direction of its layers.

The resistance opposed in a downward direction, by this intense horizontal strain, together with the weight of these protruded strata, will perhaps often be sufficient to prevent the extravasation of the crystalline matter below, and to give the predominance once more to the repressive force. The protruded laminar strata or schists will then entirely cover and conceal the crystalline axis, like a mantle. (See fig. 28. p. 195.)

In other cases a central ridge or axis of unstratified crystalline rock may be elevated in a solid state, and appear intruded like a vast dyke between the replicated strata on either side; (fig. 29.)

The crystalline mass, thus exposed by its protrusion through the overlying strata, will generally be preserved from liquefaction by the immense pressure of the column of water above it, when the process takes place at the bottom of the ocean. Where the external pressure is not sufficient to prevent ebullition, a superficial intumescense and extravasation will take place, until the temperature is by the process of evaporation so far lowered that the repressive force predominates. The liquefied crystalline matter may overspread the edges and surface of some of the overlying or protruded strata, and thus give rise to the appearance of secondary granites, syenites, porphyries, &c.

Again, portions of the expanding crystalline rock will accompany the protrusion of the lamellar strata, moving along with them, and will thus be included between the folds of the lowest of the strata, for it is obvious that the formation of any convex flexure (such as F represented in fig. 31.) by

Fig. 31.

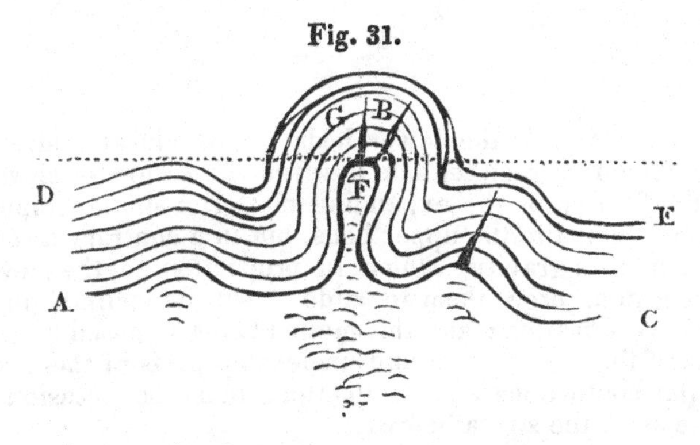

two nearly opposite forces acting on the strata in the directions A F, C F, will have the effect of reducing the pressure on the heated crystalline rock or lava immediately below, and cause it to intumesce with violence and occupy the concavity of the flexure. If the elevated strata are subsequently denuded down to the dotted line D E, the crystalline rock F will show itself between the two corresponding series of strata, and in this manner more or less narrow beds of granite, or other unstratified crystalline rock, may be expected often to occur interbedded amongst the replications of the laminar and schistose rocks, particularly of the former, which are the lowest of the series.

If, as must often have occurred, a fracture takes place at the apex of the flexure G B, the intumescent matter will be injected into the cracks, and produce one or more dykes of greater or less magnitude, according to the size of the crevices.

If the fracture is so complete, owing to the nature of the

strata, that an opening is made entirely across them, as in fig. 32, the same circumstances take place with regard to the protrusion of the crystalline axis of this *secondary* ridge, as in the case of the primary ridge, or *axis of elevation*; (fig. 29.)

Fig. 32.

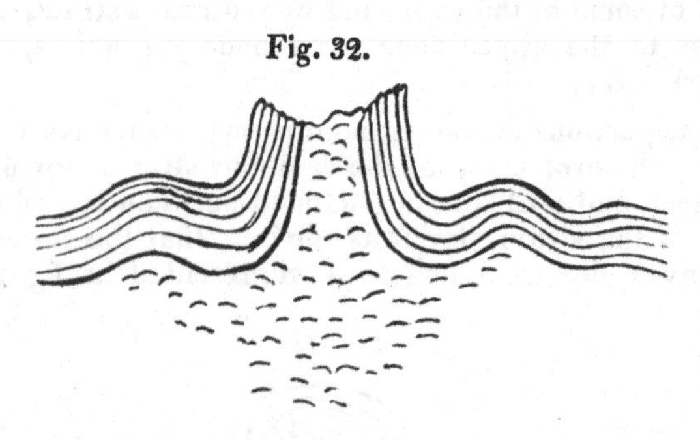

§. 3. But while the lower beds of the elevated strata are thus forced from either side towards the fissure of elevation, by the friction of the expanding matter, a similar impulse is communicated to the upper beds, but in a contrary direction, by their own gravity; which, in proportion to the elevation they sustain, urges them to slide down the inclined plane of their stratification; and this motion being opposed by the inertia of the more distant and unelevated parts of these strata, similar contortions and replications must be occasioned in these as in the strata below.

Thus there is produced a reciprocal movement in the elevated strata, the lowest beds being forcibly urged towards and through the fissure of elevation, and the upper receding from it; both movements giving rise to fractures, foldings, and other irregularities of position in the strata influenced by them, proportionate to the quantum of motion they receive, the angle to which they are elevated, and their structure, flexibility, degree of induration, &c.

§. 4. When the higher parts of these replicated strata have been worn down and carried off by the denuding forces (to which we shall presently see that they are exposed) a traveller passing across their edges will meet with repeated alternations of the same series of strata, A, B, C, D, E, &c.

When the denudation has not proceeded to the depth of the flexure of the lower beds, E, F, &c. one or more of these will be wanting, and thus the series will on some points appear to recur with great uniformity, at others some of its members will be absent, and those irregularities of this kind will be produced, which are so frequently observable in the formations of highly inclined strata composing every mountain range.

§. 5. When any of the strata are too nearly indurated to suffer such complete replication without rupture, numerous cracks and fissures will be broken through them, and particularly at the points of extreme flexure.

Many of these when they open inwardly will be filled by the intumescence of crystalline matter from below, and thus veins, dykes, or intruded masses of unstratified crystalline rocks, will appear penetrating and occasionally interbedded amongst the series of elevated strata, (syenites, porphyries, serpentines, and traps of the primitive transition and secondary formations; granite veins, in mica, and clay-slate, &c.)

§. 6. Where some of the strata are yet more completely consolidated, and therefore break much more readily than they bend, numerous vast fracture chasms will be forced across them, and the intervening portions of strata may often be elevated together with the mass of crystalline rock, (see *c*, fig. 29.) and left in an isolated position upon its summit or flanks. In this manner were probably raised those colossol and insulated pyramidal masses of dolomite which rise from the great porphyry district of the Tyrol. (See Von Buch's letter to Humboldt.)

In this locality it appears that the subjacent crystalline mass partially intumesced on some points, and intruded itself through some of the fractures, so as even to overlap the upper edge of one of the insulated dolomitic portions.

The rock is a large-grained granite in some places, and passes into serpentine on others, and finally into basalt, or augitic greenstone, which is its predominant character.*

The smaller dykes which traverse the neighbouring limestone strata are all of basalt (Porphyre pyroxenique of De Buch), and the dolomitic pyramids are apparently portions of these strata, altered by the intense heat and sublimations of the crystalline rocks with which they are in contact.

M. de Buch is of opinion that it is the intumescence and rise of the basalt (porphyre pyroxenique) that elevated the dolomite, red sandstone, shell limestone, and other secondary beds, not only of this locality but of the whole chain of the

---

* These transitions are sometimes so abrupt that I procured specimens which are granite on one side, and fine-grained greenstone on the other.

Alps. I differ from him, inasmuch as I conceive the intumescence and rise of the basalt, in this and other similar spots, to be not the *cause*, but the *result* of the elevation of these overlying strata.

A general fact, noticed by M. de Buch himself, proves this most thoroughly; viz. that wherever the basalt appears, the strata are invariably found dipping rapidly *towards* it, which is wholly inexplicable under the idea that the rise of the basalt elevated them. (M. de Buch indeed confesses his inability to explain this appearance.) If, however, we suppose the expansion of the subterranean bed of crystalline rock to have taken place at a great depth, elevating the overlying strata irregularly along the line of various fissures, as, for example, at A and B, it is clear that such fissures will open *outwardly*;

Fig. 33.

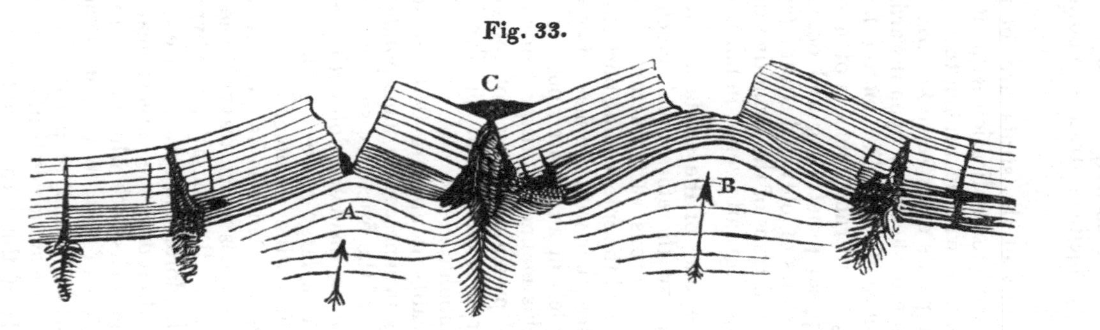

but in the interval between any two such fissures as C, another must be formed, opening, on the contrary, *downwards*, that is, towards the confined and heated lava; which, in consequence, must intumesce and fill the space afforded to it; and perhaps force its way through some minor cleft upon the external surface of the elevated rocks. One of M. de Buch's own Alpine Sections will illustrate the justness of this remark most clearly.—See Plate I, fig. 33.

If we suppose the *fissures of elevation* to have been formed along the lines of the valleys of Fassa, Eysach, and St. Pellegrin, the elevation of the overlying rocks, viz. porphyry, red sandstone, and limestone, along those lines, will have given birth to the valleys by forcing the fissures to open outwardly, while their lower extremities remained firmly closed. At the same time, by this action, in the intervening spaces, corresponding and parallel fissures must have been formed, but widening towards the lower, and closed, or nearly so, at their upper extremities. The lava, swelling up through these latter, will have occasioned the alteration of the limestone strata to dolomite, and must have forced its way exactly into those places where we find it, viz. in the intervals between the valleys, while the elevated strata must necessarily dip towards it, as was observed above to be in fact universally the case.

An inspection of the accompanying section,* which, as far as the geognostical facts go, is copied exactly from De Buch, will show how simply and easily their extraordinary appearances are reconciled and explained by this supposition, while they are wholly inexplicable under the idea of De Buch, that the protrusion of the basalt elevated the strata.

Indeed all primary vents of eruption seem to have been produced in a somewhat similar manner, viz. by the widening of a fissure towards the interior of the subterranean lava-bed, as appears from the following considerations.

The direction of the fissure of elevation created by the first successful effort of the subterranean expansive force, obviously determines that of the mountain range protruded through it. The general similarity in the arrangement of these chains, as viewed upon the surface of the globe, to that of the linear trains of volcanic vents, and the so frequent parallelism of one or more of the latter to the proximate range of mountains, is thus obviously accounted for; since the fissures of both kinds, viz. firstly, those of *elevation*, which are created by the sudden expansion of the heated lava immediately beneath, and through which the crystalline rocks and highly inclined schistose strata that form the axes of mountain chains, have been forcibly protruded; and, secondly, those of *eruption*, through which the confined caloric, and a part of the subterranean lava bed effected their escape with less disturbance and difficulty, were often produced at the same moment, and through the same continuous strata.

---

* See pl. I. fig. 33.

If, indeed, we bear in mind the general relations of position, which have been described above as subsisting between the linear trains of volcanic vents and the axes of the neighbouring mountain chains, or the outlines of the nearest continental elevations, it will appear highly probable, that the original fissures on which these eruptions have taken place, were torn through the prolongation of the lower beds of these elevated strata, by the force with which they were urged towards the fissure of elevation.

For instance, in the diagram given above (figs. 29, 30,) the lower series of strata, *a a a*, being driven by the friction of the expanding matter below, towards the orifice of the protrusion *c*, one or more wide solutions of continuity must be formed in the prolongation of these strata, as at *d d*, and in a direction at right angles to the strain, and therefore parallel to the fissure of elevation. These fractures will be probably often continued across the upper series of strata, though they will be there in great part, or entirely, choked again by the subsidence of the elevated portions of these strata, in obedience to their gravitating force. At any rate, the lava-bed immediately below must intumesce and force its way up these fractures, as far as it can penetrate (producing dykes and injected veins); and break out into absolute volcanic eruption on every point that is not effectively sealed against its expansive efforts.

At every fresh successful effort of the subterranean expansive force, by which the neighbouring continent is still further elevated, this fissure of eruption is probably opened still wider, and perhaps prolonged, and fresh vents are formed upon it.* Should, however, this cleft become so far sealed up by the abundance of crystalline matter produced upon and within it, as to offer more resistance to the disjunctive force than some unaffected part of the continuous strata, a new fissure of eruption is created, at more or less distance from, and parallel to, the former, productive of the same phenomena.

Thus, in the formation and enlargement of fissures of eruption, two forces combine, viz. the one a disjunctive tearing strain, acting in a horizontal direction, and occasioned by portions of the same continuous crust, being dragged away towards some distant fissure; the other a heaving force, act-

* At the time of the earthquake of Lima, in 1746, four volcanos broke out in eruption in one night. A similar circumstance occurred during the earthquake of Chili, in 1822.—Ulloa Voyage en Amerique. Caldcleugh's Travels in South America.

ing from below upwards, and consisting of the expansive power of the intensely heated lava immediately beneath.*

It is probable, that the first of these circumstances is the principal cause of the great difference between the phenomena that take place from these fissures and those of *elevation*, since, in the former, the intumescence and extravasation of the confined lava through the opening thus created, is opposed only by the external pressure, atmospheric or oceanic. In the latter, in addition to these elements of the repressive force, the escape of the lava is opposed by the weight and tenacity of the overlying and heaved-up strata, and, above all, by the immense resistance created in the force with which their lower beds are jammed together at the opening, by the friction of the expanding mass pressing towards it from opposite sides.

§. 7. It appears then, that of the solutions of continuity which accompanied the process of elevation, whether in the replicated strata composing the mountain, or in their more distant prolongation, the deepest and widest occasioned extravasations of the inferior lava-bed; and these remain as dykes or intruded beds, penetrating the strata, and often connected with an *overlying* bed of crystalline rock of a similar nature.

Others, which were too narrow and intricate to allow the escape of any intumescent matter in a liquid form, were yet permeable to the vapours and metallic sublimations that rose from this subjacent mass, and were partly filled by these, partly by rubbish from their sides and from above, thus giving occasion to *mineral veins*.

Those fissures which did not open downwards towards the lava were filled, in part or altogether, by rubbish alone; and these are the faults and slips of miners.

The formation of calcareous and other breccias, and veined marbles, is accounted for by the smallest of these fractures. The still unconsolidated juices of the rock oozing into its cracks and crevices, and filling them with a deposit of *finer* matter.

§. 8. Where two elevated ranges were produced at no

* During the great earthquake of Chili, in November, 1822, which was felt along the line of the Cordilleras in an extent of nearly 1000 miles, the earth was raised *en masse*; but a horizontal movement is also described to have taken place (Caldcleugh, 11, 48). The accounts of many other violent earthquakes represent such a horizontal movement to have been felt, accompanied by, and clearly distinguished from, a vertical motion, or heaving of the earth under the feet.

great distance from each other, the intervening mass of strata may have been raised to a considerable height without having its generally horizontality much disturbed, and consequently without many flexures or solutions of continuity.

This appears to have been the case with the flœtz strata composing a great part of Germany and Dalmatia, and which intervene between the Alps on one side, and the Vosges, Black Forest, and mountains of Bohemia and Hungary, on the other.

The extensive districts of flœtz strata which constitute the basin of the Mississippi and its confluents, between the Rocky mountains and the Alleghany chain, may be conjectured to offer a parallel instance.

But in such cases, the horizontality and non-disturbance of the strata occupying the middle space which separates two chains, must be generally compensated by their extreme disturbance and replication in the immediate vicinity of either range.

§. 9. Whether the elevated strata are replicated into numerous foldings on each side of the protruded ridge of crystalline rocks, a mode of procedure common it would seem to the lamellar and schistose strata, particularly the micaceous (clay and mica slates, gneiss, &c.), which appear to have shared in the peculiar flexibility that characterizes their predominant mineral ingredient); or are separated by wide solutions of continuity into detached masses, as seems to have most frequently happened to the compact limestone strata; it is obvious how either process tends to the production of longitudinal furrows, or *valleys*, on either side of, and more or less parallel to, the axis of the elevated range. In the former case,* the strata composing the sides of these valleys will appear to dip towards them on either slope—(trough-shaped). In the latter,† they will be cut off by precipitous escarpments, in which the abrupt edges of the strata will often correspond on either side.

In the chain of the Alps, the valley of the Inn, and that of the Isère, between Conflans and Grenoble, are examples of the last kind of longitudinal valley.

Those of the Adda, of Sion, of the Eisach, above Brixen, and the system of the Jura range, of the first.

Even the great valley of Switzerland, which divides the

* Sec *a*, fig. 29. p. 201. † Sec *b*, fig. 29.

Alps from the Jura, and again that of the Po, to the south of the same colossal range, appear to belong to this class of trough-shaped longitudinal valleys, or *concave flexures*. The elevated secondary strata which clothe the base of the Alpine range, both on the north and south, dip under the tertiary and alluvial beds of the plains of Lombardy and Switzerland, to rise again at the opposite side in the northern Appenine and the Jura; these two mountain ranges having all the appearance of a series of irregular convex flexures occasioned by the subsidence of their calcareous strata, in a semi-consolidated state, from off the inclined plane of the higher Alps.

The basin of the Adriatic seems to be the continuation of the valley of the Po; separating the parallel ranges of elevated secondary strata of the Appenines and Illyria.

The change in its direction corresponds to that of the primitive crystalline axis of the Alps, which is probably continued after its sudden bend to the south in the maritime Alps, and in the ranges of Corsica, Sardinia, Sicily, and Calabria.

With regard to the elevated primitive (so called) plateau of central France, the vallies of the Rhone, Loire, and Allier, present similar longitudinal troughs, parallel to the direction of the strata. There is no difficulty in supposing the extensive strata of transition and secondary rocks which probably once covered this district, to have slid away on all sides, but chiefly into the depths of the Atlantic and Mediterranean. Where their subsidence was checked by the range of the Pyrenees, they accumulated into the repeated flexures and sinuosities which characterize the secondary and transition formations of the north side of that chain, and which reach entirely to its summit, cloaking over and concealing on that side the crystalline axis of the range.

§ 10. We are thus led to recal the distinction, already established, between the primary axis of elevation along which the overlying strata were burst open and elevated solely by the successful development of the subterranean expansive force, and those secondary axes of elevation which consist in the convex flexures produced on either side of, and more or less distant from, this primary axis, by the replication of the strata elevated by the sudden development of subterranean expansion.*

---

* These secondary flexures are somewhat analogous to the waves produced by dropping a long pole into still water.

Whatever expansions took place in the inferior crystalline mass beneath these secondary convexities, were occasioned by the reduction of pressure upon it, not by the absolute increase of its expansive force, as in the primary axis. (See § 15.)

These secondary anticlinal ridges will be more or less nearly parallel to the primary ridge or axis of elevation. Where the resistances to either the protrusion or subsidence of the replicated strata are locally unequal, whether from differences in their structure and composition, or the occurrence of neighbouring obstacles, such as proximate ranges of elevation, &c. proportionate local irregularities will be produced in the direction of the secondary flexures.

The intervals between these secondary ridges, that is, the *concave* flexures, form what are so frequently met with, and which we have mentioned above as trough-shaped longitudinal vallies.

In the north of Scotland such vallies, separated by intervening secondary ridges, are numerous and remarkable, forming the basins of the greater number of her Lakes and Astuaries.

In England the basins of the Severn, Trent, &c. &c. and even of the Thames, offer similar examples; for it is probable that the different convex swellings of the flœtz strata of this country, were not occasioned by the action of any local expansion immediately beneath them, but are flexures produced by the resistances opposed to the subsidence of these strata towards the German Ocean, off the elevated range of Devon, Wales, Cumberland, and Scotland.

The local irregularity of the resistances will account for the changes of direction in the axes of these elevated ridges, which become less and less constantly parallel to the primary axis of elevation, as their distance from it increases.

These flexures present a remarkable diversity of character on different points. On some, the upper remaining beds still cover over the whole breadth of the convexity, as in the instance of the great chalk protuberance of Wilts, Berks, and Hampshire. While on other points of the same flexure or *anticlinal ridge*, the upper beds have in part disappeared, leaving a longitudinal valley, at the bottom of which the inferior beds shew themselves. The great valley of the wealds of Kent and Sussex, included between the chalk escarpments of the North and South Downs, offers a familiar example of this occurrence.

The cause of this is obvious. In the former case the

upper chalk beds possessed sufficient tenacity to sustain their elevation and curvature without splitting; in the latter a longitudinal crack opened across them, parallel to the axis of elevation. The chalk resting on beds of clayey marl (the upper strata of the greensand formation) slipped away on either side from the axis, leaving bare the lower strata of greensand. Again, the partial subsidence of this formation upon the slippery beds of Weald Clay, disclosed in turn the ironsand, which forms the visible axis of this ridge.*

The valley of Kennett offers a similar instance on a smaller scale; that of Pewsey another. In the neighbourhood of Weymouth an opening of this character in the oolitic limestone series, has discovered a small anticlinal ridge of red sandstone at its bottom.

It is indeed remarkable that these longitudinal vallies of elevation and subsidence occur most frequently in *limestone* districts; which confirms the general fact remarked above, that while the micaceous or schistose strata have usually *bent* before the elevating force, the calcareous strata have generally preferred to *break.*†

Where such anticlinal vallies are on a large scale, as the instance of the Weald of Kent, *transverse* fractures appear to have been also formed in the subsiding strata, but these will have remained more or less narrow crevice-like gorges; since no subsidence can take place away from them, that is in a direction coincident with the axis of elevation. These transverse chinks or gaps are frequently the channels, through which the drainage of the interior valley has been since effected, while they have been enlarged, and have lost the angular roughnesses of their fracture edges, by denudation, or meteoric abrasion. Such are the gaps in the chalk es-

---

* If this is the true origin of these vallies, it is obvious how improperly they are designated by the term vallies of *denudation.* Valley of *elevation and subsidence*, or *anticlinal valley*, would be perhaps a more appropriate appellation. That these districts have suffered much subsequent denudation, may be, and is, no doubt, true; but the cause of their peculiar form lies in the elevation, fracture, and subsidence of the strata which once covered this space.

† Hence these numerous cracks and fissures, for which all limestone districts are remarkable, and through which their superficial drainage is often effected. When a running stream flows for a great length of time through any series of such fissures, it wears away and scoops out the sides, giving rise to *the caverns* so frequent in these fissured limestones, and which, after the stream had been led to take another course, were often tenanted by animals, the remains of which afford the most interesting speculations to the theoretical geologist.

carpments of the Kent and Sussex valley, through which flow the Wey, Mole, and Medway, towards the north, and the Arun, Adur, Ouse, and Cockway, towards the south.

§ 11. Some of the transverse vallies also of mountain chains owe their origin in all probability to solutions of continuity, which must occasionally have been broken *across* the superficial strata, in consequence of the irregularity and violence of their elevation and subsidence. This origin is particularly applicable to those deep chasm-like recesses which contain lakes at the foot of the higher chains—e. g. the basins of the Maggiore, Lugano, Como, Iseo, and Garda Lakes, on the south of the Alps; and those of Switzerland and Tyrol on the north. The vallies of Aoste, of the Adige, the Piave, and those of the Durance, Drac, and Arc, as well as many other minor vallies having the same relative direction, viz. at right angles to the axis of elevation, probably owe their origin to the same circumstance.

The waters of the ocean in their rapid and violent retreat from the surfaces elevated above its level, would naturally occupy these fissures, and still further enlarge and deepen them, leaving vast alluvial accumulations of detritus on the plains into which the lower extremity of such gorges usually open, and where the velocity of the debacle was first checked. (Diluvium of Switzerland, Piedmont, and the Italian Lakes, &c.)

Many other transverse vallies, however, were no doubt originally scooped out by these retiring waters *alone*, without the previous existence of any directing fissure; the waters excavating their channels along those lines into which they were directed by accidents of level in the neighbouring surfaces, and the greater or less resistance of the component rocks.

The vallies of either kind have been subsequently enlarged and otherwise modified; and many others, perhaps indeed a far greater number were wholly and entirely excavated by the slow but constant and powerful action of the same causes which are still continually in force; amongst which the fall of water from the sky, and its abrasive power as it flows over the surface of the land from a higher to a lower level, is the principal.

There is good reason to conclude on grounds which will be alleged below, that the quantity of water circulating in this manner in given times, has gradually diminished from the earliest ages of the world to the present. We need not therefore be prevented from attributing to its erosive agency effects greatly exceeding in magnitude those of which it appears capable at this moment.

The evidence confirmatory of this opinion, which I am aware differs from that most generally received at the present day, will be developed at a future opportunity, and perhaps in another work. One proof, however, and that a convincing one, may be briefly mentioned here, of the *slowness* of the process by which many very considerable river vallies were excavated—viz. their *sinuosity*. This character can only be accounted for by a *lent* and *gradual* abrasion, and where this exists it is idle to talk of sudden catastrophes, debacles, or deluges, as having been the excavating forces.

It is not my intention, at present, to investigate in detail the concordance of these theoretical ideas with the peculiarities of structure of individual mountain ranges or elevated districts; but I think it must be conceded by those who have studied the physical constitution of any number of these massive protuberances with which the surfaces of our continents are wrinkled, that their general and characteristic features are simply and satisfactorily accounted for by the mode of elevation which is here attributed to them.

The accompanying ideal section of a range of mountains will illustrate their general features more fully than words can do, and exhibit a few of the accidental and irregular variations of structure which local circumstances must have frequently occasioned in them, according to the laws which have been here laid down.*

§ 12. With regard to the periods at which the several continental masses acquired their present elevation, we must conclude from the analogy of the volcanic phenomena, and from what we have recognised of the laws which regulate the development of the expansive force, that they were raised by expansive shocks succeeding one another at greater or less intervals.

Of these the greater number were probably of minor violence, similar to the earthquakes which still continue to elevate parts of the globe's surface by scarcely perceptible increments, and analogous to the minor eruptions of a volcano in the phase of a permanent or of prolonged and moderate activity. But it is also probable that a concurrence of local circumstances favourable to a long predominance of the re-

---

* See plate I.

pressive force, will have occasionally brought on a crisis of intense subterranean dilatation, a *paroxysmal expansion*, the effect of which on the solid crust of the earth will have been proportionately violent and extensive.

If it be true, as has been asserted, that outliers of the plastic clay and chalk formations have been recognised on the summits of the Savoy and Julian Alps, (and the highly inclined strata of the Nagelflue in Switzerland and Italy corroborate the opinion,) it would seem that the elevation of this colossal European chain, (and perhaps therefore of the whole of Europe) from below the level of the sea, took place by some sudden and tremendous catastrophe of this nature, at a comparatively recent geological epoch.*

The traces of (so called) *Diluvian* action will, in this case, be the result of the denuding force of the waters retreating from this elevated surface, and accompanying their retreat with frequent successive oscillatory movements.

If so stupendous a chain was raised in reality at once to its present height, the commotion necessarily produced in the ocean by such a change, will be fully sufficient to account for all the appearances of an extraordinarily violent action of water subsequent to the deposition of the plastic clay, which are visible over the whole continent of Europe.†

---

* Amongst the evidence of the recent elevation of stratified rocks we may mention the limestone cliffs of Palermo, and the whole north coast of Sicily, of Calabria, and the Promontory of Circe, which appears in many places perforated by the *Mytilus lithophagus*, at a height sometimes of upwards of a hundred feet above the level of the sea. Parts of the shells still remaining in the holes remove all doubt as to their nature.

Brocchi. Bibl. Italiana passim.
Cat. rag. p. 84.

† That the sea does share, and that in a very sensible degree, even in the ordinary commotions of the earth, was proved by the immense wave which, during the earthquake of Lisbon, swept over the lower part of that town; rose to a height of sixty feet above high-water mark at Cadiz, and along the whole coast of Portugal, and was felt even on the coasts of England and Norway. Again, at the time of the earthquake of Peru in 1746, the sea, which retreated at first, returned with fury in a tremendous wave, and overwhelmed the whole town of Callao. A similar circumstance accompanied the earthquake of Messina. The osillatory movement communicated to the sea drove an immense wave over the whole point of the Faro, which, in its return, drowned all the population of the town of Scilla on the opposite coast of Calabria.

If elevations of but a few inches or feet of vertical height produced oscillatory movements in the ocean of such violence, what must be the effect of the sudden elevation of a mountain range like that of the Alps, from the bottom of the sea?

The boulders of Jura and the southern slope of the Alps, the filling up of the valley of the Po, and the great alluvial flats of Russia, Poland, Prussia, Denmark, North of Germany, and Holland, will date from this catastrophe; while the creation and earliest outbreakings of the volcanic fissures of France, Germany, Hungary, and Italy, may be supposed to have accompanied the same event.

What subterranean expansions have taken place since that epoch, beneath Europe, have probably been of very inferior violence, and perhaps consisted rather in a general and progressive dilatation by which the mountain ranges, consisting of knots of rocks already protruded, and a great extent of the continuous stratified crust on one or both sides of them, were slowly raised by almost insensible shocks.

The geology of the other continents is not sufficiently known to authorize a conjecture as to whether the relief of their grand mountain ranges was projected contemporaneously with that of the Alps, or at a more remote or more recent period.*

But the numerous remains of terrestial vegetation, which are found in the secondary strata up to the old red sandstone, and even in some of the earlier transition rocks, prove that much land was elevated above the sea level even at these remote periods.

It is possible that paroxysmal expansions similar to that to which the Alpine chain appears to be owing, took place during these earlier ages of the globe's history; and indeed in the formation of the old red sandstone it is perhaps possible to trace the result of such a catastrophe. Other elevations may have been more gradual and less violent.

The frequent circumstance of the non-conformable portion of strata, some having been deposited across the abrupt edges of earlier and highly inclined beds, attest the occasional repetition of such phenomena.

---

* If future observations should lead to the opinion that the elevation of the principal mountain ranges of the globe was simultaneous, we must conclude this general expansion of the subterranean granite to have been produced by some sudden diminution of pressure acting generally on the surface of the earth; such as would be occasioned by the near approach of an erratic planetary body, or *comet*. If it be probable, as has been shown above (page 60), that the slight diminutions of superficial pressure, which are amongst the habitual atmospheric phenomena, may determine the production of an earthquake or local elevation of strata, what may not have been the effect of the attractive influence on the atmosphere, ocean, and earth, of a comet passing within a short distance of the globe's surface?

§ 13. It is indeed what we must conclude, from à priori evidence, that these occurrences were far more frequent and powerful in the earlier ages of the globe than they can be at present.

For it is clear that the sum of subterranean expansion taking place in fixed times, must have progressively diminished in a rapid ratio.

1stly. In consequence of the continued decrease of the expansive force, as the globe gradually cooled down, and its temperature became more nearly equalized.

2dly. From the continual increase of the repressive force, by the additions that were made in the products of volcanos, incrusting springs, &c. to the accumulation of mineral deposits on the surface of the globe, and perhaps to the body of water and atmospheric fluids which envelop that surface.

From these reasons the ratio of expansion, and, consequently, the frequency and violence of its results, whether consisting in elevations *en masse* of the solid crust of the globe, or volcanic eruptions and exhalations, must have gradually decreased from the earliest epoch of the history of the globe to the present day.

§ 14. The difference of mineral character in those crystalline rocks, whose protrusion from below accompanied the elevation of the stratified rocks, is a fact open to the same remarks which have been already made on the varieties of the rocks which were expelled from a volcanic vent, and will probably remain involved in obscurity until the researches of chemistry shall have made us more intimately acquainted with the laws by which the crystallization of different minerals is determined.

It seems probable that ordinary granite, composed of felspar, quartz, and mica, was the original or mother rock, composing what has been spoken of as the general subterranean bed of heated crystalline rock, or *lava.*

Circumstances accompanying its intumescence and reconsolidation, may be supposed in some cases to have converted the mica into hornblende, producing syenite.

A great degree of comminution, occasioned by the friction of the crystalline particles on one another, may have sometimes reduced the granite to a porphyry; small particles of felspar alone remaining visible in an apparently homogeneous base. A still further subdivision, either accompanied, or not, by some changes in the combinations of the elementary particles, may have given rise to compact felspar, (eurite or weisstein,) or serpentine; and the recrystallization of this latter rock, under favourable circumstances, to diallage rock.

The extreme disintegration of syenitic granite, will, in the same manner, have produced greenstone, and perhaps the later traps.

There are not perhaps any two of these varieties of crystalline rocks which have not been found in nature passing into each other either by sudden or gradual transitions.

It cannot therefore be deemed a rash conjecture to suppose them all to have derived from the same original; and it certainly appears most probable that the alterations they have undergone were the result of the circumstances attending their rise and protrusion towards the surface of the globe; since we have to guide us in this supposition the exact analogy of the congenerous crystalline rocks produced under our eyes by subterranean expansion, from volcanic vents, in which similar changes of mineral characters indisputably take place during the processes of emission and consolidation.

CHAP. XI.

Origin of the Strata composing the surface of the Globe, involving a Theory of the Earth.

§ 1. The explanation which has been attempted in the foregoing chapter, of the most striking differences of level observable in the solid crust of the globe, presupposes a certain degree of regularity in its structure, prior to the elevation of any of its parts above the level of the sea, since it must have been composed of concentric coats, consisting of

I. The crystalline or granitoidal matter; confined at an intense heat, by the overlying strata, viz.

II. The series of laminar and schistose crystalline rocks, viz. gneiss, mica, talc, and chlorite-schists, &c. &c.

III. The transition series, both of schistose and stratified rocks, arenaceous and sedimental.

IV. The secondary series.

} or such parts of them as were already deposited at the epochs of elevation of the different continents.

The original formation of these different beds is a question whose discussion perhaps does not properly belong to this work. A few theoretical opinions on this subject, may, however, be briefly hazarded, since they are naturally suggested by the phenomena we have been investigating.

With regard to the uppermost beds of this series, the plainly arenaceous or sedimental strata, their mode of formation differs probably not much from that by which very analagous deposits are produced at the present day in the depths of the ocean, and within its creeks and inlets, in inland lakes, the mouths of rivers, or on the surfaces of the vallies through which they flow; and what difference does exist may be perhaps accounted for by the following considerations, which have been partly anticipated in the foregoing chapter.

§. 2. Whenever any considerable portion of the strata of the globe was suddenly raised, in the manner above described, from beneath the ocean, the rush of the retreating waters displaced by it, must have possessed a vast abrasive force, by which its surface will have been more or less degraded and worn away, and its fragments carried down and deposited at a lower level.

But this is not all; for, by the absolute elevation of such a mass, the radius of the globe is dilated on this point; and to compensate for this, and preserve its equilibrium, a proportionate body of water will be immediately transferred to the opposite or antipodal point (by the same law which occasions the *antipodal tide* to the point at which the moon's attraction is exerted.) The shock produced by the rush of water to this spot from all sides, will cause its reflection in an opposite direction; and a series of oscillatory movements of this sort will occur, one wave following another, till the ordinary equilibrium of the ocean is finally restored.

The effect of each of these violent and successive waves will be to break up or wear away the inequalities of the rocks it sweeps over, and to deposit their detritus in sand or gravel banks, wherever the velocity of the waves is temporarily checked by the projecting relief of the submarine levels, or as its original impetus gradually slackens.

Here then we see that the abrasive passage of great bodies of water, animated by a violent impulse, over a large part of the surface of the globe, must have accompanied each successive elevation of any considerable portion of the strata.

Such phenomena are, perhaps, fully equal to account for the extensive denudations which our continents appear to have suffered, and for the origin of the larger fragmentary deposits of different ages. And such was probably the real character of those moving waters, which, under the name of deluges, cataclysms, or debacles, have been unavoidably inferred by geologists to have occurred at different successive epochs in the age of the world, from the evidence of these arenaceous strata.

It is in the longitudinal vallies of mountain chains, and wherever the first check is given to the displaced waters, that the coarsest fragments broken off from the highest elevations will have been left.

The finer detritus must have been carried to a greater distance, according to the weight of its particles, and the force and direction of the currents in which they are suspended.

Where the motion was trifling or null, the finest or sedimental deposits will have gradually subsided, and mixed with the precipitations which were taking place contemporaneously from the waters of the ocean, or with the bituminous and calcareous matter that accumulated in its tranquil recesses from the decomposition of animal and vegetable substances, and the opercula of marine molluscæ, coralline bodies, &c. As the depth of these beds encreased, the pressure sustained by those that were lowest, and the effect of the concretionary attraction of similar particles, effected their partial consolidation, and divided them into strata, which some subsequent expansion elevated in turn above the sea-level, where they lost by drainage all the water they retained, and were still further consolidated.

I do not, however, follow the opinion of the Huttonian geologists, that these strata are indurated by the heat transmitted to them from the interior of the globe.

The fact of the occurrence of indurated strata resting on clays and shales, sufficiently disproves this hypothesis.

The complete consolidation of many of these rocks certainly dates only from the period of their desiccation, and was therefore subsequent to their elevation. So soon as this change of level permitted the water of any stratum to run off, the immense pressure sustained would squeeze it out, and harden the remaining mass. The flints in chalk and other limestones consist of the fine siliceous particles of each stratum; which, while the whole was soft and pulpy, immediately after its deposition, sunk by their superior gravity to the bottom of the bed, and there agglomerated by the force of affinity into nodules, or, when very abundant, into thin layers. Shells, fragments of rock, and other extraneous bodies, have often settled in the lower part of each stratum in the same manner.

The seams of the strata in these, and, I believe, in all sedimental rocks, were produced by a similar species of elective aggregation between the particles of the predominant ingredient in the rock, exerted during the consecutive stages of their consolidation. The action of this attractive affinity has often produced a more or less crystalline texture, sometimes general to the mass of the rock, at others confined to one of its ingredients, which acts as a cement to the rest.

The more complete the process of crystallization, the more compact, solid, and coherent, is the resulting rock. The action of the crystallific forces in the pulpy mass must be favoured by the comminution of its particles; since the finer the particles, and the more nearly they are reduced to their integrant molecules, the more mobility they will enjoy when brought into close contact by pressure.

Those particles which are precipitated from chemical solution in water, are much finer than can be by any sediment produced by trituration; consequently, the greater the proportion of precipitated particles, to those that subside from mere suspension in water, the greater the facilities of crystallization. Hence, since it is a known general fact, that, amongst marine deposits of different ages, the oldest are almost invariably the most crystalline,* it is to be presumed that the proportion of the precipitated to the sedimental matter in the deposits of the ocean, has diminished gradually from first to last. This must be owing either to the waters which supply the ocean having less and less of calcareous or siliceous matter in solution than before, or to their having gradually lost the faculty of holding these mineral substances in solution; or finally, to both these causes united.

The waters of the ocean have two sources, viz.

I. That of land springs, many of which are still thermal, others impregnated with silex and carbonat of lime, &c., but in the earlier ages of the globe it is highly probable that the water percolating through the heated crystalline rocks, or produced by the condensation of their vapours, both possessed a higher temperature, and was more impregnated with mineral matter.†

II. Rain water. This again has two sources.

1st. Evaporation, by which pure water alone is carried up.

2dly. Volcanic vapours, by which numerous mineral substances, as muriate of soda, lime, and magnesia, sulphates of soda and lime, &c. are carried up in solution; and either immediately mingled with the waters of the ocean, if the vent is subaqueous, or carried down into it by the condensation of the vapours, if they rise from a subaerial mouth.

---

* Some very recent rocks of aqueous formation are extremely crystalline, as the Travertinos, Calc-tuff, &c., but these are entirely precipitations from calcaliferous land springs, &c. and therefore tend to confirm my argument.

† Indeed it is a known fact, that the quantity of calcareous matter deposited by many of the encrusting springs we are acquainted with, decreases from year to year.—See Brogniart, in Cuvier's Ossemens Fossiles, ii. 548.

But volcanic eruptions, and exhalations from the subterranean bed of heated crystalline rocks, appear to have diminished by degrees in number and quantity, since the earliest ages of the globe.*

Thus the quantity of mineral matter carried into the ocean in a state of solution from both these sources has progressively decreased.

At the same time it is extremely probable that the ocean originally possessed a much more elevated temperature than now; when the forces of subterranean expansion were in more frequent and violent activity; when the heated bed of granite was separated from the ocean by a thinner crust; and when volcanic eruptions were taking place in much greater number, and almost solely from the bottom of the sea, instead of, as at present, from subaerial vents.

The nature of the organic remains preserved in its deposits, which are more completely analagous in size and character to those of our tropical regions, the older the formation in which they occur, is a strong collateral proof of this fact. A higher temperature must also have permitted the ocean to hold more mineral substances in solution. And thus it appears that both the quantity of these substances introduced into the ocean, and its faculty of retaining them in solution, have diminished by degrees; which explains at once the correspondent diminution in the proportion of precipitated to sedimental matter in its deposits, and the less crystalline grain of the newer strata.

Again; the greater pressure sustained by the older members of the stratified formations must have assisted in facilitating the crystalline arrangement of their component particles. As the weight of superincumbent matter (of greater specific gravity than water) increased upon any sedimental bed, the water it contained must have been squeezed out and forced to exude upwards through the pores or clefts of the less compact strata. This would take place even under the sea-level, and the process of consolidation and crystalline aggregation, no doubt, continued, gradually accompanying the augmentation of the overlying beds; and was not finally completed previous to the elevation of the stratum above the sea-level, when its more absolute drainage was effected.

The occasional abrupt curvatures of the sedimental strata prove their induration to have been incomplete at the epoch

---

* Vide Chap. x. §. 11.

of their elevation, while their still more frequent fractures attest that they were not far from the limit of complete solidity.

For these reasons the frequent occurrence of a more or less crystalline texture, and of veins or crystals of quartz, or even occasional crystals of felspar, in some of the earlier conglomerate or sedimental strata, under such circumstances as prove their crystallization while the rock was undergoing consolidation, should appear no more surprising a fact than that of crystallized quartz or carbonate of lime in recent marles or sandstones.

In general, I am of opinion that the different proportions of siliceous, aluminous, calcareous, or other precipitations, and of quartzose, micaceous, calcareous, or other sediments; the necessary variations in the coarseness and fineness of the subsiding particles, according to the varying rate of motion of the currents, or eddies, from which they were deposited; and the different degree of pressure to which they have been exposed, will explain all the variations of texture, internal structure, and composition, in the sedimental and arenaceous deposits of all ages, without our recurring to the supposition of their having been materially affected by the transmission of heat from the subterranean lava bed, on any other points than those which have been brought into close contact with protruded portions of this crystalline bed, forced, in a liquefied state, through fissures of the stratified rocks, by volcanic action, (dykes, veins, &c.)

Even with respect to those early and more rare formations which are almost wholly crystalline; such as the saccharoidal limestone, quartz rock, &c. I see no reason to suppose that they owe their origin to any other cause than the abundant precipitation of carbonate of lime or silex from the primitive ocean; which was at the time perhaps *locally supersaturated;* and the subsequent consolidation of these beds, and the intense pressure to which from their position they were necessarily exposed.

The general intermixture of micaceous and other fragmentary particles in these rocks, proves them to participate in some degree in the sedimental character; though the proportion of precipitated matter generally predominated.

Where on the contrary the sedimental matter, particularly mica, was in excess, the mica, talc, and chlorite slates were produced; the quartz, which continued to be deposited, often agglomerating into nodules; and the mica partially recrystallizing: while occasionally other minerals were formed, such as asbestos, epidote, garnet, &c. &c. by the capricious play of the forces of chemical affinity.

§. 3. With regard to the more completely crystalline, and nearly granitoidal rocks, which constitute the gneiss formation, their origin is a question of more difficult solution. Their laminar and schistose structure is, however, so analogous to that of some volcanic rocks, (the clinkstones and lamellar basalts) that little doubt can be raised of their being indebted for it to a similar cause. It has been suggested as probable in a former part of this essay, (Chap. V. §. 2.) that this structure was occasioned in the latter class of rocks by their subjection, after a partial intumescence, to a resolidification, attended with an incomplete recrystallization, and produced by an intense pressure between two opposite forces; owing to which the crystals arranged themselves in such a manner as to have their shortest axes in the direction of the pressure, and their longest in a plane perpendicular to it.

Now in the case of the zone of laminar and crystalline rocks which is here supposed to have originally encompassed the globe, what more evident than that, between the repressive force of the overlying strata, and the expansive force of the subjacent lava-bed, it must have been exposed to just such an intense compression acting exactly in the direction of the radius of the globe, and therefore at right angles to the planes of the crystalline laminæ?

If then we imagine a general intumescence of an intensely heated bed of granite, forming the original surface of the globe, to have been succeeded by a period in which the predominance was acquired by the repressive force, occasioned by the condensation of the waters on its surface, and the deposition from them of various arenaceous and sedimental strata, (the transition series,) the lamellar structure of the gneiss formation is at once simply explained. This structure may have been subsequently increased by the friction of the different laminæ against one another, as they were urged forwards, in the direction of their plane surfaces, towards the orifice of protrusion, along with the expanding granite beneath; the laminæ being elongated, and the crystals forced to arrange themselves, in the direction of the movement.

In what manner this general superficial intumescence, and the succeeding phenomena, took place, is a question involving a complete theory of the earth; which, in the present imperfect state of our chemical knowledge, it is difficult to frame on a satisfactory basis.

§. 4. Perhaps, however, I may be pardoned for offering the following conjectural rough sketch of such a theory; of the imperfections of which I am fully conscious, but to which the ideas developed in the foregoing chapters necessarily lead; and which appears to me to explain the various appearances presented by the known constitution of the globe, more simply, and intelligibly, and to be more accordant with the general and constant processes of nature, than any other which the spirit of geological enquiry has yet started.

I imagine then the mass of the globe, or, at least, its external zone to a considerable depth, to have been originally, (that is, at, or before the moment, in which it assumed the position it now holds in the planetary system,) of a granitic composition; composed probably of the ordinary elements of granite; and having a very large grain; the regular crystallization having been favoured by the circumstances under which it previously took place; though, as to what those circumstances were, I do not venture to hazard a supposition.*

---

* The above theory is not incompatible with the idea of the granitic involucrum of the globe having been produced by the superficial oxydation of a metallic nucleus. This opinion, originally started under the authority of a great name, has since met with, I believe, many supporters. It must, however, be owned that it smells a little of the laboratory; and that, though by possibility it may prove to have been a good guess in the end, yet one or two chemical experiments, and one or two analogical inferences drawn from them, do not form a sufficient basis for the erection of a theory on the origin of the globe.

Volcanos have been adduced in support of this opinion, and have been supposed to proceed from the penetration of water to the metallic nucleus of the earth

This hypothesis is at variance with the whole tenor of the preceding essay; and cannot, I think, be in any way reconciled to the detail of the volcanic phenomena.

But it is equally untenable in itself—For how is it possible to suppose that water can percolate to this intensely heated nucleus, without being vaporized long before it reaches the limits of the metallic part? What force is to urge the water to this depth? Whence did this fluid originally derive? and how, above all, after having been once expelled, together with a quantity of the oxydized matters through the fissures of admission, is it to find its way again to the same point, so as to produce a repetition of the eruption?

The only evidence that can be adduced at all, in favour of such an hypothesis, consist in the circumstance that most volcanos break out from the sea or the sea-coast.

This fact, however, has been fully accounted for by proofs that it is the local absence of volcanic vents, which determines those elevations of the superficial strata on the large scale, to which our continents are owing;

I suppose that on reaching its actual orbit, and even before, it enjoyed a great diminution of the pressure which had previously crystallized, or at least preserved it in a state of crystallization, at an intense temperature, (perhaps at an integrant part of the sun ?*)

This diminution of compression produced its expansion; which must have taken place from the surface inwardly towards the centre; the outer coat being rapidly volatilized more or less, and the next zone only liquefied partially; the dilatation experienced diminishing towards the interior.

If the diminution of previous compression commenced, as is likely, before the planet reached its actual orbit, great part of the atmosphere, formed by its superficial vaporization, will have been parted with on its passage from the sun; (like that of a comet, which, in its tail, seems to be continually leaving behind part of its substance in an aeriform state.) The aeriform envelope which remained, or was evolved, after it had settled in its present position; (the centrifugal force which drove it from the sun, having reached an equilibrium with the centripetal force, or gravitation, which draws it towards the centre of the system ?) formed, and still forms, its atmosphere; and by subsequent condensation, part of its ocean, or *main reservoir* of *superficial waters.*

Supposing the globe to have had any irregular shape when detached from the sun, the vaporization of its surface, and of course, of its projecting angles, together with its rotatory motion on its axis, and the liquefaction of its outer envelope, would necessarily occasion its actual figure of an oblate spheriod. As the process of expansion proceeded in depth, the original granitic beds were first partially disaggregated, next disintegrated, and more or less liquefied; the crystals being

---

and that therefore, in the same quarter of the globe, the rise of land above the sea-level must have generally varied inversely with the development of the volcanic phenomena.

* The phenomena of Aerolites seem to indicate that masses of mineral matter of a crystalline texture, granitoidal structure, sometimes almost identical with granite in mineral composition, are still occasionally projected through the range of our planetary system, and come within the sphere of the earth's attractive influence. The coat of glossy black varnish which forms on the surface of these stones, and of their fragments as they fall, seems to be somewhat analogous to the original micaceous envelope of the globe.

merged in the elastic vehicle produced by the vaporization of the water contained between the laminæ.*

Where this fluid was produced in abundance by great dilatation, that is, in the outer and highly *disintegrated* strata, the superior specific gravity of the crystals forced it to rise upwards; and thus a great quantity of aqueous vapour was produced on the surface of the globe.

As this elastic fluid rose into outer space, its continually increasing expansion must have proportionately lowered its temperature; and in consequence a part was recondensed into water, and sunk back towards the more solid surface of the globe.

And in this manner, for a certain time, a violent reciprocation of atmospheric phenomena must have continued. Torrents of vapour rising outwardly; while equally tremendous torrents of condensed vapour, or *rain*, fell towards the earth. The accumulation of the latter on the yet unstable and unconsolidated surface of the globe, constituted the primæval ocean. The surface of this ocean was exposed to continual vaporization owing to intense heat; but this process abstracting caloric from the stratum of water below, by partially cooling it, tended to preserve the remainder in a liquid form. The ocean will have contained, both in solution and suspension, many of the matters carried upwards from the granitic bed in which the vapours, from whose condensation it proceeded, were produced, and which they have traversed in their rise. The *dissolved* matters will have been silex, carbonates, and sulphates of lime, and magnesia, muriates of soda, magnesia and lime, and those other mineral substances which water at an intense temperature and under such circumstances was enabled to hold in solution. The *suspended* substances will have been all the lighter and finer particles of the upper beds where the disintegration had been extreme; and particularly their mica, which, owing to the tenuity of its plate-

---

* If indeed this fluid was not itself produced by the forced union, at an intense heat and pressure, of hydrogen and oxygen gasses, generated by the volatilization of some of the mineral particles composing the outer laminæ of the crystals; by which exactly the same effect would be produced, as if the water had been mechanically enclosed between these laminæ. Though this conjecture is by no means warranted by the present state of chemical knowledge, it does not I believe contradict any known fact, and would explain many problematical circumstances; particularly the non-occurrence of water in the older granites.

shaped crystals, would be most readily carried up by the ascending fluid, and will have remained longest in suspension.

But as the torrents of vapour, holding these various matters in solution and suspension, were forced upwards, the greater part of the disintegrated crystals by degrees subsided; those of felspar and quartz first, the mica being, as observed above, from the form of its plates, of peculiar buoyancy, and therefore held longest in suspension.

The crystals of felspar and quartz, as they subsided, together with a small proportion of mica, would naturally arrange themselves so as to have their longest dimensions more or less parallel to the surface on which they rest; and this parallelism would be subsequently increased, as we shall see hereafter, by the pressure these beds sustained between the weight of the supported column of matter and the expansive force beneath them. These beds I conceive, when consolidated, to constitute the gneiss formation.*

The further the process of expansion proceeded in depth, the more was the column of liquid matter lengthened, which gravitating towards the centre of the globe, tended to check any further expansion. It is therefore obvious that after the globe settled in its actual orbit, and thenceforward lost little more of its enveloping matter, the whole of which began from that moment to gravitate towards its centre, the progress of expansion inwardly would continually decrease in rapidity; and a moment must have at length arrived when the forces of expansion and repression had reached an equilibrium, and the process was stopped from progressing further inwardly by the great pressure of the gravitating column of liquid.

This column may be considered as consisting of different strata, though the passage from one extremity of complete solidity to the other of complete expansion, in reality must have been perfectly gradual. The lowest stratum, immediately above the extreme limit of expansion, will have been granite barely *disaggregated*, and rendered imperfectly liquid by the partial vaporization of its contained water.

---

* The marks of mechanical attrition in gneiss are well known; its broken crystals, fragmentary nodules, and passages into quartz-rock, and other more decidedly mechanical deposits. It is also a general remark that the largest grained gneiss lies nearest to granite, and that it becomes finer, more slaty, and more micaceous, as it recedes from the granite. (Jameson's Geognosy, p. 115.)

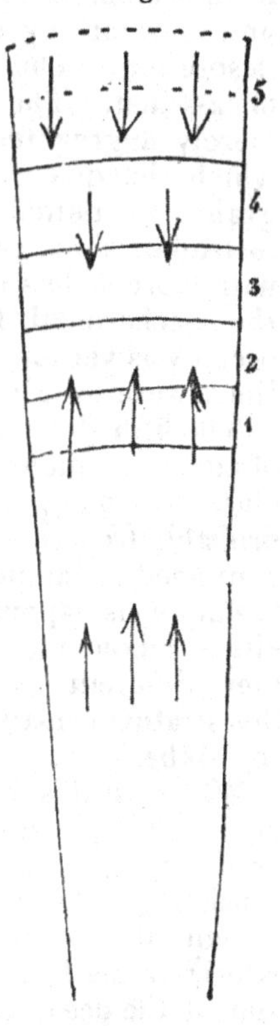

Fig. 34.

The second stratum was granite *disintegrated*; aqueous vapour having been produced in such abundance as to be enabled to rise upwards, partially disintegrating the crystals of felspar and mica, and superficially dissolving those of quartz. This mass would reconsolidate into granite, though of a smaller grain than the preceding rock.

The third stratum was so far disintegrated that the greater part of the mica had been carried up by the escaping vapour *in suspension*, and that of quartz *in solution*; the felspar crystals, with the remaining quartz and mica, *subsiding* by their specific gravity, and arranging themselves in horizontal planes.

The consolidation of this stratum produced the gneiss formation.

The fourth zone will have been composed of the ocean of turbid and heated water holding mica, &c. in suspension, and quartz, carbonate of lime, &c. in solution, and continually traversed by reciprocating bodies of heated water rising from below, and of cold fluid sinking from the surface, by reason of their different specific gravities.

The disturbance thus occasioned will have long retarded the deposition of the suspended particles. But this must by degrees have taken place, the quartz grains, and the larger and coarser plates of mica, subsiding first, and the finest last.*

But the fragments of quartz and mica were not deposited alone; a great proportion of the quartz held in *solution* must

* It is acknowledged by geologists that the large grained and most quartzose varieties of mica-slate are the oldest, (that is, the lowest of the series,) and that the rock becomes finer and approaches more nearly to clay slate, as it recedes from the supporting gneiss.

Mica-slate also exhibits marks of a more completely mechanical deposition than gneiss. (Jameson p. 119.)

have been precipitated at the same time as the water cooled, and therefore by degrees lost its faculty of holding so much in solution. Thus was gradually produced the formation of mica-schist. The mica imperfectly recrystallizing, or being merely aggregated together in horizontal plates, between which the quartz either spread itself generally in minute grains, or united into crystalline nucleis. On other spots, instead of silex, carbonate of lime was precipitated, together with more or less of the micaceous sediment, and gave rise to the saccharoidal limestones. At a later period, when the ocean was yet further cooled down, rock salt, and sulphate of lime, were locally precipitated in a similar mode.

The fifth stratum was aeriform, and consisted in great part of aqueous vapour; the remainder being a compound of other elastic fluids (permanent gasses,) which had been formed probably from the volatilization of some of the substances contained in the primitive granite, and carried upwards with the aqueous vapour from below. These gasses will have been either mixed together, or otherwise disposed, according to their different specific gravities, or chemical affinities; and this stratum constituted the atmosphere, or aerial envelope of the globe.

When, in this manner, the general and positive expansion of the globe, occasioned by the sudden reduction of external pressure, had ceased (in consequence of the *repressive force* consisting of the weight of its fluid envelope, having reached an equilibrium with the *expansive force*, consisting of the caloric of the heated nucleus), the rapid superficial evaporation of the ocean continued; and, by gradually reducing its temperature, occasioned the precipitation of a proportionate quantity of the minerals it held in solution, particularly its silex. These substances falling to the bottom, accompanied by a large proportion of the matters held in suspension, particularly the mica, in consequence of the greater comparative tranquillity of the ocean, agglomerated there into more or less compact beds of rock (the mica-schist formation), producing the first crust or solid envelope of the globe. Upon this, other stratified rocks, composed sometimes of a mixture, sometimes of an alternation, of precipitations, sediments, and occasionally of conglomerates, were by degrees deposited; giving rise to the *transition* formations.

Beneath this crust a new process now commenced. The outer zones of crystalline matter having been suddenly refrigerated by the rapid vaporization and partial escape of the water they contained, abstracted caloric from the intensely heated nucleus of the globe. These crystalline zones were of unequal density, the expansion they had suffered diminishing from above downwards.

Their expansive force was however equal at all points, their temperature bearing every where an inverse ratio to their density. But when, by the accession of caloric from the inner and unliquefied nucleus, the temperature, and consequently the expansive force, of the lower strata of dilated crystalline matter was augmented, it acted upon the upper and more liquefied strata. These, being prevented from yielding *outwardly* by the tenacity and weight of the solid involucrum of precipitated and sedimental deposits which overspread them, sustained a pressure out of proportion to their expansive force, and were in consequence proportionately condensed, and by the continuance of the process, where the overlying strata were sufficiently resistant, finally consolidated.

This process of consolidation must have progressed from above downwards with the increase of the expansive force in the lower strata, commencing from the upper surface which, its temperature being lowest, offered the least resistance to the force of compression.

By this process the upper zone of crystalline matter, which had intumesced so far as to allow of the escape of its aqueous vapour, and of much of its mica and quartz, was resolidified; the component crystals arranging themselves in planes perpendicular to the direction of the pressure by which the mass was consolidated, that is, to the radius of the globe. The gneiss formation, as already observed, was the result.

The inferior zone of barely disintegrated granite from which only a part of the steam and quartz, and none of the mica, had escaped, reconsolidated in a confused or granitoidal manner; but exhibits marks of the process it has undergone, in its broken crystals of felspar and mica; its rounded, and superficially dissolved grains of quartz; its imbedded fragments (broken from the more solid parts of the mass, as it rose, and enveloped by the softer parts); its concretionary nodules, and new minerals, &c.*

* Humboldt asserts the *oldest granites* (that is the *lowest*) to contain most quartz and least mica; the increase of mica creating a foliated structure by the pallets being placed in a parallel position, and then the rock passes into gneiss, or foliated granite; the older granites distinguish themselves also by their large grain, and the regular crystallization of the constituent minerals; by the absence of included nodules, masses of a different grain,

Beneath this, the granite, which had been simply disaggregated, was again solidified, and returned in all respects to its former condition. The temperature, however, and with it the expansive force of this inferior zone, was continually on the increase; the caloric of the interior of the globe still endeavouring to put itself in equilibrio, by passing off towards the less intensely heated crust.

This continually increasing expansive force must at length have overcome the resistance opposed by the tenacity and weight of the overlying consolidated strata. It is reasonable to suppose that this result took place contemporaneously, or nearly so, on many spots; wherever accidental circumstances in the texture or composition of the oceanic deposits, led them to yield most readily; and in this manner were produced those original fissures in the primæval crust of the earth, through some of which (fissures of elevation) were protruded portions of the inferior crystalline zones in a solid or nearly solid state, together with more or less of the intumescent granite, in the manner above described; while others (fissures of eruption) gave rise to extravasations of the heated crystalline matter, in the form of lavas, that is, still further liquefied, by the greater comparative reduction of the pressure they endured.

The protrusion of the foliated rocks, gneiss, mica-schist, clay-slate, &c. was chiefly occasioned, as was observed above, by their peculiar structure; the parallel plane surfaces of their component crystals, particularly the plates of mica, sliding with facility over one another; while the laminar structure of these rocks was in turn increased during this process, the crystals being elongated in the direction of the motion, as in the case of the clinkstones and pearlstones of the trachytic formation.

The violence with which these rocks were elevated, and

---

of heterogeneous beds (grunstein and limestone) of accidental minerals, such as epidote, hornblende, steatite, garnets, tin, oligistic iron, actinolite, &c. &c.; and by a *non-porphyritic* structure. The newest granites are characterised by the presence of tin, magnetic iron, hornblende, diallage, garnet, talc, and chlorite replacing the mica, ***and by a tendency to pass into graphic granite!***—*Essai sur la Classification des Roches.*

The increase of mica in the new granites is, I think, only ***apparent***, the tabular plates composing the hexahedral prisms of the older rock having been separated and broken up by the friction accompanying the process of intumescence, so as to be disseminated more generally through the rock, and consequently to make more show, although perhaps really in lesser quantity.

The same circumstance is very apparent in some varieties of trachyte.

the replications into which it forced them, produced numerous fractures which were filled in one of the following modes, viz.

I. By intrusion of the intumescent granite from below, when the crack opened directly towards it.

II. By rapid expansion of the crystalline rock composing the sides of the fissure, when its temperature was so great as to produce the ebullition of the contained water, at the diminished pressure occasioned by the creation of the fissure. The matter produced in either manner, would be, probably, a fine-grained granitoidal rock; or by the continuance of disintegration, a *porphyry* or *serpentine:* if its particles were comminuted to an extreme degree by the friction attendant on its intumescence, they may subsequently, during the process of slow consolidation, have affected entirely new combinations, and given rise to diallage-rock, syenite, hornblende rock, &c.

III. Where the temperature of the rock was not sufficiently high to produce rapid ebullition in the walls of the fissure, the aqueous vapour alone which remained within the rock would be forced to *exude* from either side, by the pressure of the mass above, and occupy the cavity, bringing with it the silex it holds in solution, which, as it gradually cooled, will have crystallized; sometimes together with other mineral and metallic matters sublimed from the lower part of the fissure, where the temperature was higher. Hence the numerous quartz veins of gneiss, but particularly of mica-schist, and which are so frequently metalliferous. Whenever any of these fissures opened outwardly into the ocean, or any of the looser and more permeable strata, the immense pressure acting on the lower beds will have propelled the vapour upwards through these

---

* According to Dr. Macculloch the gneiss of the Hebrides is always locally dislocated, bent, and contorted, in proportion to the abundance of granite veins which intersect it. Sometimes the gneiss is *shifted*, at others not, by these veins. In such granitic contemporaneous veins, particularly in those which traverse granite, very regular quartz crystals often occur in cavities (*pockets*, *geodes*), that is air-bubbles, formed by the accumulation of the highly compressed aqueous vapour into large volumes; the silex having been subsequently precipitated on the sides of the cavity by slow degrees, as the water cooled. Hence too the extreme rarity of the gaseous fluid which occupies these cavities before they are broken into.

adyts, and produced the continuance of its exudation from the walls of the vein, together with fresh mineral matter in solution; part of which crystallized on the sides of the fissure, and the surfaces of what fragments occurred in it.

The disturbance of the water occasioned by this process may account, perhaps, for the clay and bouldered fragments, which are sometimes found in mineral veins; though it is more probable that these have found their way into open fissures from above.

The quartz veins of clay-slate, quartz rock, &c. were, most probably, formed, in the same manner, by exudation of water, with silex in solution, from the sides of cracks, produced in them, whether by external force, or by the contraction accompanying their desiccation. In the case of gneiss or granite, if the rock was not completely consolidated when the rupture took place, and yet at too low a temperature for ebullition, the crystalline matter of the sides may have been squeezed into the fissure together with the exuding water; and here the solid crystals of felspar and mica enjoying a greater facility of motion in this aqueous vehicle, will have reunited by degrees into larger crystals. The felspar appears sometimes to have forced the quartz, which was precipitated from the water as it cooled, to arrange itself in plates between the parallel laminæ of its reaggregated crystals, so as to produce what is called *graphic granite*,* which usually occurs in the granitic veins of the gneiss formation.

The formation of the *pure* quartz veins will always have been subsequent to that of the granitic; since the former require the previous refrigeration of the rock, as well as its partial consolidation; and this accords with the numerous known examples of granitic veins traversed by quartz, while the contrary fact is, I believe, unknown.

These sudden partial elevations of the crust of the globe, and the other various causes which at this period disturbed

* It is possible that in the glassy felspar crystals of trachyte, which are almost pure silex, though they retain the crystalline form of felspar, plates of silex may be in this manner appliquès in thin laminæ between the different crystalline molecules, and that the general absence of grains or crystals of quartz in this recent granitoidal rock may be thus accounted for. The soda, which is also an element of these crystals, and by the help of which the quartz was probably taken into solution by the aqueous juices of the rock, augments the evidence in favour of this supposition.

the tranquillity of the primitive ocean, produced violent waves and currents, which broke up and triturated the projecting eminences of its bottom, and distributed their fragments in alluvial conglomerate strata, wherever the turbulence of these moving waters was partially checked. The surface of the globe at this period consisted chiefly of mica-schist; and hence mica and granular quartz predominate greatly in the conglomerates of this epoch (greywackè and granular quartz rock.)

The subsidence of these fragments was accompanied and succeeded, during intervals of tranquillity, by that of the finer micaceous particles suspended in the ocean, and by the precipitation of silex or carbonate of lime; and hence the occasional passages of greywackè into mica-slate, fine-grained quartz-rock, and transition limestone. Hence too the almost uniformly dark colours of the two latter rocks, when of too fine grain for their elements to be recognized.* The protrusion of crystalline rocks from below, through fissures of eruption successively formed during the deposition of these earliest arenaceous and sedimental strata, accounts for the frequent occurrence of intruded dykes, veins, &c. of syenites, porphyries, serpentines, and trap-rocks, in this formation. These different crystalline rocks being, as I conceive, merely varieties of granite disintegrated in a greater or less degree, and *recrystallized* under peculiar circumstances, which occasioned the formation of new mineral combinations. In this manner, epochs of intense local development of subterranean expansion, producing partial elevations of the earth's crust, and local depositions of conglomerate rocks, will have, probably, often alternated (according to the general laws of these phenomena) with periods of comparative tranquillity, during which the sediments and precipitations of the ocean were of a finer grain; and hence the alternations of the various arenaceous and sedimentary strata which compose the secondary formations.†

But in the mean time the temperature of the ocean being somewhat lowered, it began to be thickly peopled with orga-

---

* The bitumen resulting from decomposed animal and vegetable matter aided in producing this colour.

† For the same reason that mica predominates in the earliest fragmentary deposits, the micaceous schists having been first exposed to the denuding forces, clay and quartz began subsequently to prevail when the excavations had reached to the subjacent formation of gneiss and granite. Where the attrition of these fragments was long continued, the mica and felspar will have been triturated so completely as to remain for a long time in suspension; while the rounded grains of quartz alone by their superior hardness

nic beings of simple structure, and a constitution suited to the circumstances under which they were created. Portions also of the elevated strata having risen into the vicinity of, and even above the level of the sea, the vegetable world sprung into life; at first plants of the simplest characters, and fitted to live under water, or in the overloaded and heated atmosphere, were alone produced. Their entwined masses rotting on the surface and sides of the stagnant bays and gulfs where they grew, or into which they had floated down from inland lakes, were slowly carbonized; and subsiding to the bottom were, at last, interbedded with strata of water-worn fragments, brought into the sea by the copious fluvatile torrents; which, in that age, must have been on a scale of magnitude proportioned to the rains which poured down with scarcely intermittent violence on the first-born continents. (The Coal strata.)

At length as the temperature of the globe's surface, and consequently of the ocean and atmosphere, gradually diminished (by radiation of caloric from the atmosphere into the surrounding void) the quantity of water taken into circulation through the atmosphere decreased; the continents were no longer deluged by perpetual floods of rain.* New tribes of

---

having partially resisted the attrition were deposited by themselves in the pure quartzose sand or sandstone strata, which are of frequent occurrence in the secondary and tertiary beds. The clay subsided probably afterwards, either by itself, so as to form *shales*, or in conjunction with carbonate of lime, in the coarse marly limestones.

* The following are the principal causes which must have occasioned a much more rapid and abundant circulation of water through the atmosphere, and over the surface of the globe, in the earlier ages of its history.

I. An inferior atmospheric pressure, owing to which the rapidity of superficial evaporation was proportionately great.

II. Greater abundance of aqueous vapour directly exhaled from the interior of the earth, through every practicable pore of its crust, and by means of volcanic spiracles.

III. The facility of evaporation resulting from the heated state of the earth's crust, which probably long retained all the water which was in contact with it, far above the boiling point under the pressure of the atmosphere alone. The *surface* of the ocean was probably for a very long period in continual ebullition; the water heated by contact with the solid bottom rising from its superior specific gravity. The earliest zoophytes and organised inhabitants of the sea were probably able to support this high temperature. We know that some plants and animals will live and flourish in thermal springs of a very elevated temperature.

organized beings, both vegetable and animal, came into existence—the marine deposits contained by degrees less of precipitated matter, and consequently became less crystalline and compact; strata of shales, dull limestones, marls, and chalks, succeeding those of clay-slate, marbles, and breccias; until the succession of ages finally brought about the condition in which the globe at present exists.

In consequence of the general nature of this gradual change, wrought by the slow refrigeration of the outer zones of the globe, the congenerous rock formations of any individual period have usually some characteristic features by which they resemble one another, and differ from those of preceding or subsequent ages. This is particularly the case with regard to the organic remains they enclose. These distinguishing marks, however, should not lead us into the error of supposing that all similar and *coeval* deposits were strictly *contemporaneous;* much less that they were uniformly deposited over the whole surface of the globe.

No geologist questions the fact, that the various freshwater formations which have been already noticed in England, France, Germany, Italy, Hungary, and Spain, are *local* formations in the strictest sense produced in separate basins or districts of more or less extent. Yet the analogies which exist between the formations of these different localities are striking, and, to say the least, *as complete* as those subsisting between the more ancient marine formations of different localities. The resemblances therefore of these early formations in different countries does not in the least tend to prove that they are relics of the same continuous strata. Because we find in the south of Russia, for example, a rock formation resembling the chalk of England or France, there is not the slightest reason for supposing similar deposits to have been ever formed in the intermediate space.

The same thing is true of every other marine formation, and particularly of the alluvial or arenaceous strata, which were more necessarily *local* deposits, than the sedimentary beds; since the heavier drifted fragments must have been deposited from the marine currents at the first check which their velocity experienced. The same current which deposited a coarse conglomerate at the foot of the masses of rock it was wearing away, perhaps left the materials of a sandstone at a greater distance; farther on, a *shale* or a *marl;* and still farther, a clayey limestone. When the progress of the excavating process, or other circumstances, by altering the relief of the subaqueous surface, had changed the direction and conduct of the currents, the nature of the deposits may have been locally altered; gravel or sand being drifted upon those points where the finer clayey and calcareous particles had formerly subsided, or where coralline zoophytes and molluscous animals had lived and died in myriads; and these being in turn left upon the previously formed arenaceous beds. In this manner limestones, shales, and sandstones, were, no doubt, produced contemporaneously on different spots; and many alternations of such strata may have taken place within no very lengthened period. It is therefore evidently needless to imagine unprecedented and extraordinary changes general to the whole surface of the globe for the sake of explaining such alternations which, no doubt, are from time to time taking place in an analogous manner at the bottom of the actual ocean.

Indeed, with regard to the formation of the successive conglomerate or arenaceous strata, as well as of the traces of excavation and denudation visible on the dry surfaces of the earth (both of which appearances are evidently connected in their origin; the fragmentary strata being the accumulated wreck of the rocks excavated and worn away), it certainly appears to me at once the most reasonable supposition à priori, and the best warranted conclusion from the facts within the scope of our observations, that the superficial destruction of the irregular protuberances of the earth's crust, by the erosive force of water in motion, has gone hand in hand with the accumulation of their fragments in alluvial strata, ever since the epoch of production of the first conglomerate rocks; that it has proceeded generally by a lent and uniform process, gradually diminishing in energy from the beginning to the present day; but occasionally presenting partial crises of excessive turbulence, resulting from accidental combinations of circumstances favourable to the maximum of violence; and particularly the sudden elevation of continental masses—that this process is, for the most part, the same which *still* operates in the circulation of water, through the atmosphere and ocean, and over the surface of the earth; and finally that the intermittent occurrence of circumstances productive of an excess of disturbance and abrasive energy, as well as the gradual diminution of intensity in the general process, are both of them suppositions warranted by what we already know of the laws which regulate the circulation of water, and of the constitution and active subterranean forces of our planet. Whereas, if this explanation be rejected, we must have recourse to the gratuitous invention of vague and unexampled occurrences, referable to no known law of nature; but which under the specious names of deluges, cataclysms, convulsions, &c. serve merely as convenient cloaks to our ignorance, and solve the difficulty only by the magic of an empty sound.

§. 5. It appears to me therefore on the whole, that the formation of the grand mineral masses of every age composing the known crust of the globe, is attributable to *three* primary modes of production, distinct in their nature, but of which the products have been often confused and mingled together, from circumstances of isochronism or collocation.—These are,

I. The chemical precipitation of various mineral substances, but particularly silex and carbonate of lime, from a state of solution in the ocean, or other body of water; as its temperature and solvent powers gradually decreased.

II. The subsidence of particles of mineral matter, of various degrees of coarseness, from a state of suspension in the ocean or other reservoir, into which they had been taken up, either by the violent escape of aqueous vapour from the interior of the globe, by the abrasive force of marine and fluviatile currents, or finally by the decomposition of the shells of molluscous animals, which possessed the faculty of elaborating their coverings from the substances they procured from sea-water.

III. The elevation of crystalline matter through fissures in the crust of the globe, which had been already formed in the two former modes; this rise being occasioned either by the expansion of a lower bed, in which case the rock was elevated nearly in a solid state; or by its own intumescence, owing to a sudden diminution of compression; in which case the matter rose in an imperfectly liquid state, and at a high temperature.

All the characteristic differences observable in the successive formations of every kind, may, I conceive, be satisfactorily traced to the gradual diminution in frequency and energy of the productive causes, the varying nature of the original materials acted on, and the chemical and mechanical changes they have undergone during the process; and with this consideration in view these three modes of production are perhaps fully equal to account for the origin of all the great mineral masses observable on the surface of the globe. They have also one immense advantage over most, perhaps over all, of the hypotheses that have as yet been brought forward to explain the same appearances; and which speaks volumes in their favour; and this is, that *they are all still in operation*,—with diminished energy, it is true; but this is the necessary result of their nature.

The *first* mode still gives rise to calcareous and siliceous rocks of great solidity, and even of a crystalline texture, in the vicinity of certain thermal or mineral springs.

The *second* still produces strata of marls, sand, and gravel, at the bottom of the sea, of inland lakes, and in the beds of rivers; which strata bear a very decided analogy to the earlier sandstones and limestones.

The *third* is in constant operation wherever volcanos break out into activity, or earthquakes produce elevations of the solid strata.

It has hitherto been a serious impediment to the progress of knowledge, that in investigating the origin or causes of natural productions, recourse has generally been had to the examination, both by experiment and reasoning, of what *might be* rather than *what is*.

The laws or processes of nature we have every reason to believe invariable. Their *results* from time to time vary, according to the combinations of influential circumstances; but the processes remain the same. Like the poet or the painter, the chemist may, and no doubt often does, create combinations which nature never produced; and the possibility of such and such processes giving birth to such and such results, is no proof whatever that they were ever in natural operation.

Another fertile cause of error in most of the various geological theories which have been successively presented to the public, lies in the notion (originating in our love for the marvellous) that every appearance out of the common course of things is owing to some one novel and extraordinary *cause*. This disposition it is which has at various times originated so many false, foolish, and mischievous opinions, as to the causes of events, which the progress of civilization and knowledge has discovered to belong to the ordinary, and never ceasing laws of nature. Hence in early ages every striking coincidence was attributed to the interference of a deity called fortune, or by some other name—every remarkable action was the effect of magic—every circumstance which could not be accounted for by the limited knowledge of the day, a miracle.

Hence too the term, *supernatural*, applied to the imaginary causes of any extraordinary event. As if any thing could occur that was not caused by some law of nature; or as if we have any right to suppose that these can suffer interruption from any ulterior cause.

It is this wonderworking spirit, whose tendency leads to the imagining unexampled and hypothetical causes to account for every appearance which is but a little removed from those commonly submitted to our observation, that appears to have inspired the inventors of most of the theories hitherto published, to account for the origin of the appearances presented by the mineral constitution of our planet.

But for this unfortunate propensity, a better method of reasoning from effects up to causes, might have been adopted, than has been yet generally introduced into the science of geology; or rather observation would have been more frequently substituted for reasoning.

The theory of the globe, which I have hazarded above, and which, I am aware, requires much ulterior development, and perhaps some corrections, to render it generally satisfactory; consists simply in the application of those modes of operation which nature still employs, on a large scale, in the production of fresh mineral masses on the surface of the earth, to explain the origin of those which we find there already.

If, after fair discussion, and with all reasonable allowances, it is found adequate to this purpose, its truth will be established on the soundest possible basis—the same upon which rests the whole fabric of our knowledge on every subject whatsoever, the supposition namely, that the laws of nature do not vary, but that similar results always are, have been, and will be produced, by similar preceding circumstances.

[illegible] events which the progress of [illegible] knowledge [illegible] belong to the ordinary, and never ceasing [illegible] nature. Hence in early ages every striking [illegible] [illegible] to the interference of a deity called [illegible] or by some other name [illegible] remarkable action was the effect of [illegible] which could not be accounted for by the limited [illegible] of the [illegible].

[illegible] could [illegible] [illegible] of the [illegible] or as if we [illegible] to suppose [illegible] interruption [illegible].

[illegible] whose tendency leads to the [illegible] causes to account for every appearance [illegible] little removed from those [illegible] appears to have [illegible] pub-lished [illegible] the origin of the appearances presented [illegible] mineral [illegible] of our planet.

[illegible] unfortunate [illegible] up to [illegible] adopted, [illegible] been [illegible] of geology [illegible] would have [illegible] more [illegible] substituted [illegible].

The theory of the globe, which I [illegible] hazarded above, and which [illegible] requires [illegible] development, and perhaps [illegible] generally [illegible] tory, consists [illegible] of [illegible] the same [illegible] on a large scale, in the [illegible] on the surface of the earth [illegible] there [illegible].

[illegible] its truth will be esta-blished [illegible] upon which [illegible] the whole fabric of [illegible] subject [illegible] the laws of nature [illegible] are, have been, and will be produced [illegible] similar [illegible] circumstances.

APPENDIX, No. 1.

List of known Volcanos in Recent or Habitual Activity.

VOLCANOS OF EUROPE AND THE ADJACENT ISLES.

Vesuvius. (Kingdom of Naples.) First recorded eruption, A.D. 79. by which Herculaneum, Pompeii, and Stabiæ were buried.

Before this epoch it is probable that the eruptions took place from the central crater of Somma, which formed a single conical mountain. This volcano has since experienced a great diversity of phases, and during a period of nearly two centuries, viz. from 1109 to 1306, remained in a state of complete inactivity. The crater at this time contained woods and a few small lakes.

Again there was a century of absolute repose after the year 1538. This intermittence was terminated by the violent paroxysmal eruption of 1631.

In 1760 eruptions burst out at once from 15 different points of a fissure broken from the summit to the base of the mountain—each of these openings vomited lava and scoriæ. The lavas of this volcano have long consisted solely of leucitic basalt; but from the quantity of pumice in the conglomerate strata of Somma, it is probable that they were formerly of a felspathose or trachytic composition. The extinct vents of he Campi Phlegræi in the immediate vicinity of Vesuvius having almost uniformly produced trachytic lavas, corroborate this opinion.

The *Monte Nuovo* in the Bay of Baiæ having been thrown up by an eruption in the midst of the Campi Phlegræi, so lately as the year 1538, must be counted as the site of a volcanic vent recently in activity. A considerable warmth is still felt at the bottom of the crater, and very thin vapours escape from some of its crevices.

The neighbouring crater of the Solfatara is supposed also to have been in eruption in the beginning of the 12th century.

Ætna. (Sicily.) A volcanic mountain of great size and of considerable regularity. It has been habitually active, at least from the earliest historical ages. Its slopes are sprinkled with above 70 parasitical cones, the product of lateral explosions. The lavas of this volcano are of a felspathose basalt, in some cases passing into greystone, from the abundance of felspar. They present little variety.

Stromboli. (Lipari isles.) A small and singularly interesting volcano from the permanence of its phenomena. Its lavas are chiefly a very black, dense, and heavy augitic basalt.

There are, however, rocks in the island of a very felspathose character.

Volcano, (also one of the Lipari isles.) Its known epochs of eruption are in the years 1444, (when large fragments are said to have been projected to a distance of six miles;) 1550, 1739, 1775, 1780, and 1786. The supramarine portion of this volcanic mountain exhibits a minor cone with a central crater, (now in the condition of a Solfatara) rising from the cavity of an earlier and very extensive crater, hollowed out by some paroxysmal eruption from a conical mountain of proportionate dimensions. (See p. 158. fig. 24.)

The lavas of Volcano are trachytic, and some of the later currents are completely vitrified. (obsidian.)

All the other Lipari islands are volcanic, but no eruptions are known to have taken place from them within the reach of history. On some points, however, of Lipari itself, there still exist hot springs, and emanations of vapours highly charged with mineral matter. The Campo Bianco, a mountain at the eastern extremity of Lipari, is a large cone entirely composed of currents of highly vitreous vesicular obsidian, interbedded with pumice.

Ischia,—having broken out in eruption so lately as the fourteenth century, must be included in the list of still active volcanos. In the earliest historical times it appears to have been subject to frequent and extremely violent phenomena, which more than once destroyed or drove away all the settlers who had been attracted to the island by its extreme fertility, and delicious climate. In figure this island presents a very regular volcanic mountain studded with parasitical cones. Hot springs and sulphureous vapours rise upon different points. Its lavas are very various and beautiful in appearance. All extremely felspathose; greystone, and approaching to trachyte. Many are highly porphyritic; some present knots of pure felspar crystals as large as the fist. Others are marbled, brecciated, &c. with a mixture of lavas of different mineral composition, colour, and grain. The tufas (or felspathose conglomerates) with which this island abounds are remarkable for their greenish tints which derive probably from the abundance of augite in the lavas.

Santorini, in the Grecian Archipelago, was in eruption in the year 1707. The many smaller islands and rocks which have at different known periods been thrown up in the vicinity of this principal isle, may be considered as parasitical eminences of the same submarine volcanic mountain, and their phenomena as proceeding from lateral and subsidiary vents of the same fundamental volcano.

Milo, though the epochs of its eruptions are unknown, is a volcano of recent aspect, having a very active solfatara in its central crater; and many sources of boiling water and steam.

Iceland contains numerous volcanic mountains, of which the following are habitually active, that is, have repeatedly broken out within record; viz.

Hecla, whose last eruption dates from 1766.

Krabla.

Kattlagiaa, which after a quiescence of 64 years burst out violently in July 1823; the previous paroxysmal eruption of 1755 was still more terrible; immense torrents of water caused by the melting of the snows, rushed from the mountain and deluged the country—it was accompanied by electrical phenomena, by which cattle were killed, &c. The eruption lated a year.

Eyafialla Jokul, which after a similar intermittence of more than a century's duration, produced a violent eruption in December, 1821. The explosions lasted till June, 1822, when the mountain opened at its base, and gave vent to an immense current of lava.

Eyrefa Jokul. This vent has been tranquil since 1720.

Grimvatn. In 1716 an eruption burst forth from a lake of this name, probably the crater left by some earlier paroxysmal explosion by which the former volcanic cone had been wholly blown into the air.

Skaptaa Jokul and Skaptaa Syfsel. Two neighbouring volcanos, whence in 1783 a prodigiously violent series of eruptive phenomena took place, and ravaged a vast extent of country around. The lava issued from three sources in the plain, at the base of the mountains, distant about eight miles from one another, and spreading in concentric waves deluged a space of more than 1200 square miles in extent. The pulverulent ejections which terminated this eruption, lasted a twelvemonth; during which, the whole atmosphere of Iceland was continually darkened by thick clouds of ashes.

Wester Jokul, burst out in 1823, but produced no lava current.

In the year 1000 an eruption occurred from the Guldbringe Sysel, a mountain which has been tranquil ever since; another in 1340 near Reckianes.

In 1583 an eruption was observed at a great distance in the sea; and during the phenomena of Skaptaa

Jokul in 1783, the summit of a cone was raised above the sea level by submarine explosions, at a distance of thirty miles from the coast.

Thus the number of volcanos which have been recently in activity in Iceland amounts to eleven or twelve. Of these Hecla alone has been for any long period in the phase of permanent or very frequent activity. Thirteen eruptions are recorded of this volcano since the year 1137. The last was in 1766; since that period it has remained inactive; and as this last eruption was preceded by a quiescence of seventy-three years, the volcano may be presumed to have passed into the phase of long intermittences.

The whole island of Iceland appears to be of volcanic formation, and every mountain a volcano, either active, or temporarily extinct. The island may in fact be conceived as a great crust of rocks, both fragmentary and solid, which has caked above the mouth of a vast subterranean cauldron of heated lava. The steam evolved through the fissures of this crust, gives rise to the numerous hot springs for which Iceland is noted.

Esk. A volcano in the Isle of John Mayen, off the Coast of Greenland, was seen in eruption in April, 1818. Projections of ashes took place every three or four minutes, and reached a height of 4500 feet. It appears that there are other volcanos on the Coast of Greenland, or at least that there have been at an earlier epoch.

Volcanos of the African Isles.

No active volcanos have been recognized in the continent of Africa, but the islands which border upon it in either ocean are almost exclusively volcanic, and offer many vents habitually eruptive.

The Azores are uniformly of volcanic constitution. *St. Michael*, the largest island of the group, suffered severely from earthquakes in 1810—11, until in February of this last year a submarine eruption burst forth two miles from the coast, and left a shoal on which the sea breaks. On the 13th of June, the same year, after other earthquakes, another island was thrown up two miles and a half beyond the first. This cone contained a crater 500 feet in diameter; it was 300 feet high above the sea. This island being solely composed of fragmentary ejections has been gradually worn away by the action of the waves and currents, and has at length been reduced to a shoal below the water-level. In 1628 a similar isle was produced, and disappeared from the same cause, between Terceira and San Miguel; and another in the same place in 1721, which in two years time had been completely levelled. There are now eighty fathoms water above the spot where it rose.

San Miguel has various volcanic cones, but none have been in recent activity. It contains, however, a solfatara at Villa Franca; and hot springs, and emanations of sulphuretted hydrogen gas rise from divers points of the island.

El Pico, is a regular volcanic cone 9000 feet in height. It broke out in eruption in 1718, but has been tranquil ever since.

The neighbouring island of *San Georgio*, was devastated in 1808, by an eruption from the centre of the island.

The island of *Fayal* has also a very regular crater in its centre, which was in activity in the year 1672.

The lavas of the Azores are chiefly trachytic.

The Canaries are likewise solely of volcanic origin.

The Peak of Teneriffe is amongst these the most celebrated as being of immense size, and reaching a height of 12,140 feet. The highest peak has, however, been inactive ever since the island has been inhabited, and the eruptions of the volcano have taken place from lateral vents and chiefly from the crater of Chahorra. The last paroxysm in 1768 was preceded by a quiescence of ninety-three years. It lasted three months. According to Cordier, the scoriæ projected at this time occupied from twelve to fifteen seconds in falling from their extreme height to the ground; and hence would appear to have reached an altitude of above 3000 feet.

In 1706 a current of lava, which filled the harbour of Garachico, was observed to flow to a distance of eighteen miles in six hours. The lavas of the Peak consist of trachyte.

Palma, a very regular conical mountain, produced a violent eruption from a lateral vent in the year 1558. The lava reached the sea, and by heating it killed quantities of fish. In 1646 and 1677 new vents were formed, and eruptions of considerable magnitude took place.

Lancerote. This island was the scene of the most terrific volcanic phenomena in the year 1730. It appears from the accounts collected on the spot by M. De Buch, that during the space of three years continual eruptions were taking place from numerous vents consecutively opened upon one line stretching directly across the island. A great part of its surface was deluged by lava currents, and the remainder buried under showers of scoriæ and ashes. The quantity of these fragmentary ejections has been favourable to the fertility of the island; for when the affrighted inhabitants, who had fled to the neighbouring isle of Fuertaventura, returned, they found a soil far richer than that they had left, and were enabled to cultivate the vine which they had not previously done. During this period, explosions and jets of scoriæ and smoke took place from the sea, numerous fish were killed, which floated on the surface, together with masses of pumice; and a pyramidal rock rose above the water-level, which was afterwards united to the island by the accumulation of new matter. The lavas of Lancerote are basaltic.

Ferro. A volcano on this island broke forth in the year 1677: and again produced an eruption of six week's duration in 1692.

Fuego, one of the Cape Verd Islands. The other islands of the group are equally of a volcanic constitution; as also are those of Fernando Po, and Prince's Island, in the Bight of Biafra.

Ascension island has been mentioned by some writers as an active volcano; but though entirely volcanic, and of a fresh aspect, the epochs of its eruptions are unrecorded.

Bourbon. The principal and central volcanic mountain of this island, which constitutes the greater part of its mass, has been apparently long extinct. A smaller and very regular cone rises on the south, within a circular range of cliffs, the walls of a vast and ancient crater. This volcano has been in almost constant external activity since the earliest colonization of the island. M. Hubert, who has observed its phenomena since the year 1766, asserts, that it has been in violent eruption at least twice in every year during this period. The phenomena of this interesting volcano have been dwelt on at length in the body of the above work. Its lavas are partly trachyte and partly basalt.

In Madagascar a volcano is said to exist, but we possess no correct account of its situation or conduct.

Zibbel-teir. An active volcano was seen by Bruce, in the island of this name, in the Red Sea. It had four apertures which launched forth smoke and flames. (scoriæ.)

Volcanos of America.

In *Greenland* a volcano exists, which appears to have been in eruption in June, 1783, at the same period as the Skaptar Jokul in Iceland.

One was observed by Cook at the extremity of the promontary of *Alaska* on the N. W. coast, which, with two more to the N. E. of this point, remarked by himself and by La Peyrouse in activity, form the prolongation of the volcanic chain of the Aleutian isles.

A volcano is mentioned also by La Perouse in lat. 41. north of Cape Mendocino. Five are reported to exist in *California*.

Mexico contains five, viz.

Colima, which was observed in eruption by Dampier, who describes it as having two mouths, both in activity at the same time. It is a mountain of immense bulk, and nearly 10,000 feet in height.

Popocatepec, is above 16,000 feet high, and appears to be at present in permanent activity; though it is known to have been quiescent for a long period previous to the year 1530, when it burst out into violent eruption.

Orisaba is a volcanic mountain, also above 16,000 feet in height. No eruptions have been recently remarked from this vent.

Tuxtla, S. E. of Vera Cruz, broke out in 1793. The ashes were transported as far as Perote, a distance of fifty-seven leagues in a straight line.

Xorullo. An account of the remarkable eruption of this volcano, in 1759, is given in No. 2. of the Appendix.

M. de Humboldt remarks that these volcanic vents of Mexico are ranged along a line perpendicular to the axis of the great Cordilleras. They appear therefore to have been produced from a *transverse*, instead of a longitudinal or parallel fissure.

In the Provinces of Guatimala and Nicaragua a line of active volcanic vents runs parallel to the axis of the Cordilleras. Those which have been reported as occasionally in eruption amount to twenty-one. Their names are, as given by Humboldt, Sonusco, Sacatepec, Hamilpas, Atitlan, Fuegos de Guatimala, Acatinango, Sunil, Tolima, Isalco, Sacatecoluca near the Rio del Empa, San Vicente, Traapa, Besotlen, Cocivina, Viego, Momotombo, Talica near San Leon di Nicaragua, Granada, Bombaeho, Papagallo, and Barua.

The Province of Grenada in S. America contains the volcanos of *Sotara*, *Purace*, *Pasto*, and *Rio Frugua*.

The Province of Los Pastos, those of *Cumbal*, *Chiles*, and *Azufral*.

The principal volcanos of Quito are the

Antisana, which rises above 18,000 feet above the sea. It has been tranquil since 1590.

Rucupichinca, which was in activity in 1660.

Cotopaxi, was observed in eruption by Bouguer and La Condamine in 1742. The projections of incandescent scoriæ reached a height of more than 3000 feet above the summit of the mountain.

The melting of the snow occasioned a terrific deluge which devastated the plains below, and destroyed 800 persons.

The eruptions of 1743 and 1744 were still more disastrous.

The French Savans remarked that the great paroxysmal explosion of this mountain, which happened in 1583, launched blocks of pumice containing from 300 to 350 cubic feet, to a distance of nine and ten miles.

Tunguragua, which burst out in 1641.

Sangay. This volcano has been in constant activity since the year 1728.

Chimborazo is an immense trachytic dome, which, however, has never been remarked in eruption.

Carguairazo, which in 1698 vomited a prodigious quantity of mud, or water mixed with trachytic ashes, covering a surface of eighteen square leagues with this substance which the natives call Moya.

But one active volcano, that of *Arequipa*, is known in Peru.

The volcanos of Chili are very numerous. They follow the direction of the Andes. Their eruptions have been frequently observed to coincide in time with the earthquakes by which this country is often desolated. Their names are given as the mountains of *Copiapo*, *Coquimbo*, *Choapa*, *Aconcagua*, *Santiago*, *Peteroa*, *Chillan*, *Tucapel*, *Callaqui*, *Chinal*, *Villa-rica*, *Votuco*, *Huaunauca*, *Ojorna*, *Huaiteca*, and *San Clemente*.

Volcanos in the Islands Dependent on America.

The Leeward Isles are in great part of volcanic constitution. An eruption occurred in *St. Vincent's* in 1718. It commenced with the violent shock of an earthquake, and was accompanied by a hurricane. Ashes obscured the air for a long period, and fell at a distance of 130 leagues. The detonations were heard to the same distance.

Another eruption broke forth from the same crater, which had remained since the last in the state of a souffriere, in 1812, preceded by upwards of 200 earthquake shocks felt during the course of a twelvemonth. It began by one violent explosion, projecting an immense column of ashes into the air to a great height. After four days incandescent scoriæ were first observed, and immediately afterwards the lava flowed out in streams. Earthquakes preceded the expulsion of the lava. After this had ceased to issue, the detonations continued for twelve hours diminishing gradually in violence, till they terminated entirely.

The island of Grenada contains on extinct crater, and numerous springs of boiling water, whence it may be concluded that the epoch of its eruptions is not very remote.

St. Lucia has a very active solfatara, and springs of hot water and steam.

The volcano of *Guadaloupe* was in eruption in 1797, but its phenomena were confined to the projection of ashes, pumice-stones, and sulphureous vapours.

Nevis, Montserrat, and St. Christopher's, all contain solfataras in full activity. Martinico, Dominico, and St. Eustace exhibit numerous craters, and some sources of boiling water.

The lavas of the Leeward Isles present varieties both of trachyte and basalt.

The group of the Aleutian Isles, which may be reckoned as a dependance of America, contains, it is said, six volcanos in activity; viz. Kanaga, Tatavanga, Oominga,

Oonalaska, Omnak, and Ourimak. The last burst out with violence in 1820.

The cluster of *Revillagigedo* is wholly volcanic; but no eruptions have been mentioned from them.

Trinity Isle, in lat. 56, at some distance from the coast of America, contains a volcanic mountain, which has been seen in eruption by many navigators.

The Continent of *Asia* contains no active volcanos of which any certain accounts have been collected, with the exception of those which are found in Kamskatcka.

Mount Elburus, in Persia, the highest peak of the Caucasian range, is often cited as a volcano; on what authority I know not. Another has been supposed to exist north of Irak, in the province of Khorasan.

The Mountains of Tourfan and Bisch-Balikh, which form part of the great Altai chain of Central Asia have been lately reported to send forth continually flames, smoke, and ammoniacal vapours. They are probably therefore in the state of solfatara.

The Peninsula of Kamskatcka appears to be in great part the product of volcanic eruptions. Those vents which remain in activity, are the mountains of

Awatscha; its most terrific eruption of which any notice has been preserved occured in 1737. It was accompanied by a violent earthquake, and extraordinary agitation of the sea, which invaded and inundated the land. An eruption happened in 1779, when Captain Clerk visited the coast. Another volcano is mentioned by Captain Clerk, in the same neighbourhood; and a third called

Apalskoi, has recently emitted much smoke from its summit.

Two volcanos called *Shevelutsch* and *Joupanowskaia* are mentioned by Auteroche. Another, the

Kamskatkoi Sopka, is of immense height. Since

1728 it has produced frequent eruptions of considerable magnitude. By some of these the country has been covered with ashes in a radius of 300 versts round the volcano.

The neighbouring volcano of Tolbatschink constantly smokes. It was in violent activity in 1739.

Many other conical volcanic mountains, with craters, &c. are met with in Kamskatcka, which have not, however, been recently in eruption. The whole promontory is subject to frequent earthquakes, and hot springs are common.

The *Kurile* chain of Isles is a prolongation of the volcanic range of Kamskatcka, and appears to consist of a train of volcanic mountains, of which many are still occasionally subject to eruption, viz.

Alaid, an island about twenty miles south of Cape Lopatka, which burst out in eruption in the year 1793; and has emitted smoke ever since;

Tkarma; Tshirinkutan; Rachkoke; Mutova, and *Etorpu.*

The *Japanese* Islands contain ten occasionally active volcanic vents, of which three occur in the principal isle, Niphon. Their eruptions are described by Kæmpfer as extremely violent and destructive.

Sulphur Island, in the Loo-choo Archipelago, emitted an abundance of sulphureous vapour when the Lyra, Captain Hall, passed it in 1816.

The Polynesian Archipelago, which seems to owe its existence principally to volcanic action, contains numerous spiracles in frequent activity. It is to be regretted, however, that we have not more detailed and scientific accounts of the natural phenomena and productions of this interesting quarter of the globe.

Amongst the Philippine islands, *Manilla* is said to contain numerous volcanos. Mindao one. Mindanao produced a violent eruption in 1764, which strewed the neighbouring country with fragmentary matter several feet in depth, and drove the greater number of inhabitants to emigration.

In the district of Kalagan is a volcanic mountain in the state of solfatara.

Borneo, according to many authors, contains some volcanos; but their precise situation is not known.

Barren Island, has a very active vent in continual eruption. It launches into the air to a great distance rocks of several tons weight.

The Moluccas abound with volcanos. One of them, *Sorca*, was the scene of a tremendous eruption in 1693.

The Peak of *Ternate* ejects pumice in vast quantity.

Motin was subject to a great eruption before Capt. Forrest's arrival in 1772. And a terrific commotion of the volcano of Gounapi in the isle of Banda, some years ago ravaged the whole island. Another eruption took place from the same vent in 1820, which projected fragments as large as the houses of the natives to a height equal to that of the mountain itself.

Ternate presents one active mouth. *Tidore* another.

Celebes is said to contain a great many.

Sanguir, between Mindanao and Celebes has one of the largest volcanos of the globe.

Tomboro in the Isle of Sumbawa gave vent to a terrific eruption in July, 1815. It commenced with subterranean detonations which were heard at Sumatra, 970 miles distant in a straight line, and were mistaken for the firing of musketry. Ashes were carried to Celebes and Java 300 miles off, in such quantity as to obscure the air there. The sea rose twelve feet higher than it ever did before, and a hurricane accompanied the explosions, and did much injury.

Flores, *Daumer*, and another small island between Timor and Ceram, contain each of them a volcano in occasional activity.

Java is thickly sprinkled with volcanos, which are ranged in straight lines along the island. The following is a list of their names, and the dates of their last eruptions :

Salak. ......1761....Eruption.

Tankuban. ..1804....Sulphureous vapours.

Guntur .....1807....Eruption.

Gagak ......1807....Partial eruption.

Chermai ....1805....Eruption.

Lawn.......1806....Sulphureous vapours.

Arjuna, 10,614 feet in height, emits a constant column of smoke.

Dasar. ......1804....Eruption.

Lamongan...1806....Idem.

Tashem .....1796....Idem.

Klut ....1785....Idem.

The Mountain of Galoen-gong, which had never been reputed volcanic, broke out with terrific violence in October 1822. The eruption began with a tremendous burst, which sent up a column of stones and ashes that darkened the whole sky. The lava deluged a vast extent of surface. 2000 persons were destroyed.

The Mountain called Payandayang had been one of the highest volcanos of Java, when between the 11th and 12th of August, 1772, a violent explosion blew up the whole of its substance, and replaced it by a cavity measuring fifteen miles by six.

Sumatra contains, according to Marsden, four active volcanos. It is probable that more will be hereafter recognised there. The inhabitants are alarmed when these vents are tranquil for any length of time, as they have found by experience that such intermittences of the volcanic phenomena are sure to be succeeded by violent earthquakes, from which they suffer severely.

Two volcanos were observed in eruption by Dampier in *New Guinea*, in the year 1700; and at the entrance of the straights which separate this island from New Britain is an insular volcano seen in eruption successively by Dampier, and Le Maire and Schouten, and D'Entrecausteaux.

Two volcanos were observed by Cartaret amongst the Duke of York's and Queen Charlotte isles; and two others by Forster in the group of the New Hebrides. One of them, Tanna, was seen in eruption by Cook, 1774; and by D'Entrecasteaux in 1793.

The *Marianne* isles are said to contain nine volcanos in habitual activity.

Tofooa, one of the Friendly Isles, has been seen in eruption by various navigators; as also Mouna-roa, a mountain of Owyhee in the Sandwich Isles.

Amsterdam Island is another volcano, which has been observed in agitation by all who have visited it.

Volcanic eruptions are said to have been remarked from a mountain in one of the islands lately discovered by Russian navigators, between New Georgio and Sandwich Land; as well as from another peak within Sandwich Land itself.

The volcanos mentioned in this list amount to upwards of a hundred and seventy; but the accounts that have been collected of some are very vague; and while, for the reasons stated in the early part of this work, it may be presumed that many more volcanic mountains exist in occasional activity, it is obvious that many which have been long extinct are of course liable to be again restored to activity by the combination of influential circumstances.

APPENDIX, No. 2.

On the Eruption of Jorullo in Mexico.

It has been asserted in the early part of this work, that the accounts which we possess of the phenomena of volcanic eruptions as observed in all parts of the globe by different eye witnesses, present a remarkable uniformity of character; and that, when due allowance is made for mistakes arising from the misuse or misunderstanding of terms, and for the exaggerations into which observers would be naturally led by the terrific nature of the phenomena, they are found to offer few other variations than what depend on the degree of energy displayed.

I am not, indeed, aware of any exception to this uniform concordance, but that of the remarkable phenomena which are described by M. de Humboldt, as having taken place during the eruption of the volcano of Jorullo in Mexico, in the year 1759. The singularity of this account is so great, that it will be well worth while to give up a few pages to its examination.*

Previous to 1759 it appears that the plain from which Jorullo now rises, presented traces of former volcanization; its soil being composed of tufa; and the neighbouring mountains consisting of trachyte and basalt. In September of that year, a violent series of eruptions took place, of which M. de Humboldt distinguishes the results in the following order.

---

* See Essai Geognostique, p. 351. Essai politique sur la Nouvelle Espagne, vol i. p. 251.

1st. The production of six volcanic cones, composed of scoriæ and fragmentary lava.

2dly. That of a promontory of basaltic lava proceeding from the summit of the largest of these cones (Jorullo), which still emits wreaths of vapour from the interior of its crater.

3dly. The elevation in a convex form of the plain (four square miles in superficial extent) upon which the cones were thrown up, and the centre of which is occupied by the largest (Jorullo), at whose base the plain is higher by 550 feet than without the limits of this space. The plain, which M. de Humboldt calls "un terrain bombè en forme de vessie," and the convexity of which he attributes to *inflation from below*, is represented as closely sprinkled with thousands of flattish conical hillocks, from six to nine feet high, formed of basaltic balls, separating into concentric leaves, imbedded in a *black clay*. These hillocks, as well as some large fissures which traverse the intermediate plain, act as so many *fumarole*, giving out thick clouds of aqueous vapour combined with sulphuric acid, and at a very high temperature.

The two first mentioned products of this eruption are of ordinary occurrence, and testify that at least for the greater part the volcanic action took effect here in the usual mode.

The phenomena of the 3d class are remarkable, and deserving of the greatest consideration, as appearing at first sight to differ materially from any hitherto observed, and as referred for this reason by M. de Humboldt to a mode of volcanic action invented by him for the occasion, and of which no other recorded eruption has ever afforded a parallel.

And here, with the utmost respect for the great talents of this first of scientific travellers, and giving all due weight to the impression which appears to have been made upon him on the spot, I own myself still unable to coincide in his opinion as to the mode of formation of this remarkable plain. And this for the two following reasons.

1st. In the first place, the appearances presented can be without the least difficulty explained by the ordinary mode of operation of volcanos. In which case we are bound to dismiss one so extraordinary and unparalleled as that brought forward for the purpose by M. de

Humboldt, and which, however brilliant or seducing to the imagination, it would be unwarrantable to persevere in.

2dly. *All* the supplementary arguments which M. de H. adduces, are completely invalid, and instead of supporting his theory rather tell against it, as will be proved directly.

1. What are the positive facts with which we are acquainted relative to this eruption, divested of all theoretical assumptions?

In the month of September, 1759, prodigious volcanic eruptions took place from six different openings, arranged on a single line of very little extent, upon the Mexican plateau. Their fragmentary projections produced six large volcanic cones; the central one being 1700 feet in height. A massive current of lava projects from the side of this last hill, having evidently flowed out of the crater at its summit. If any lava currents were produced by the apertures marked by the other cones, they do not shew themselves; but the plain from which these hills rise exhibits a great intumescence, or convexity of surface. Its superficial soil consists of horizontal layers of a black clay, in which augite crystals are thickly disseminated. The same clay, but enveloping concentric balls of *basalt*, composes the numerous little hollow conical protuberances, (or bubbles), with which the surface of the plain is sprinkled.

Now on comparing these appearances with those which result from ordinary volcanic eruptions, little other difference is perceivable than that the quantity of *lava* produced, or at least remaining visible, bears but a very small proportion to the violence of the eruptions, and the immense quantity of scoriæ thrown out. It seems extraordinary also that but one out of six cones should have given rise to a lava current. Hence a suspicion arises that a greater quantity of lava was in fact emitted, but that it is concealed by the sprinkling of triturated scoriæ or volcanic sand, which these large cones must have thrown out during the latter period of their eruptions.

In this manner the lava currents produced by the eruption of Vesuvius in October, 1822, lie at this moment covered to a depth of from two to ten feet by the finer fragmentary substances ejected by the volcano during the last days of its paroxysm. And what renders the analogy still more striking, these *ashes*, which, from the fineness of their comminution mixed into a retentive paste with the torrents of rain that immediately followed the eruption, present the appearance of strata of *a black clay*, precisely like those described by M. de Humboldt, as forming the surface of the plain of Malpais.

The convexity of this plain is therefore most naturally accounted for by supposing it to form the surface of a great bed of lava, resulting from the union of different currents, which, owing to the previous flatness of the surface on which they were poured forth, their simultaneous emission in great abundance from so many neighbouring orifices, and their very low degree of liquidity,* united into a sort of pool or lake of lava, which spread itself on all sides with great reluctance, and therefore would necessarily remain thickest and deepest where the lava was produced in greatest abundance, diminishing in bulk from thence towards the limits of the space it covered; i. e. would assume precisely the convexity of form peculiar to Malpais. The subsequent projections of loose and pulverulent matter from the six craters, and principally from Jorullo, will have increased that convexity, covering it with strata of volcanic ashes and augite crystals, which were reduced to the appearance of a black clay by mixture with rain water.

The fact that the only visible lava current proceeds from the *crater* of Jorullo, is a strong confirmation of this opinion; since it is at once obvious why this is seen, while those that may have been emitted *previous* to the formation of the other cones are concealed; and it becomes also probable that this *promontory* of lava is merely one extremity of the current of Jorullo, which dipping under the strata of ashes, probably unites with the streams proceeding from the other apertures to form the substratum of the whole convex plain.

Thus there is no difficulty in accounting for the convexity of the plain of Malpais by the effects of the most ordinary volcanic phenomena; let us see whether this supposition will explain the other remarkable appearances it is said to exhibit.

And here a fact recorded by M. de Humboldt himself not only tends to confirm, but may be almost said to prove, the correctness of the view I have taken of the nature of the plain. He says that in 1780 the temperature of the fissures which penetrate the surface of the plain and its hillocks, was so high, that a cigar might be lighted by plunging it to the

* The very coarse grain of the lava of Jorullo (*Dolerite*, Humboldt) warrants this assumption of its extremely imperfect liquidity.

depth of a few inches into them. Now I think it impossible to account for this without allowing the whole plain to have consisted of lava in a state of incandescence immediately beneath its outer crust; a circumstance to be expected, even so much as 20 years after its emission, in a bed of lava more than 500 feet in thickness, since Hamilton observed of one of the smaller currents of Vesuvius, that three years after its production, a stick might be inflamed by thrusting it into one of the crevices of the rock.

It remains only to account for the formation of the small hillocks (hornitos) with which the surface of the plain is thickly studded. And here I must again have recourse to the results of the highly instructive eruption which took place from Vesuvius in October 1822.

All lava currents are well known both during their progress, and for a long time, often indeed many years afterwards, to disengage torrents of aqueous and sulphureous vapour. If these are produced on any point in considerable quantity, while the superficial lava is yet soft, their expansion raises up a portion of this into a small dome or bubble, which sometimes remains entire, at others it is broken through, leaving the tattered fragments of lava that separate to give passage to the vapour, in an upright position from their sudden congelation.* This process is in fact one of the causes of the numerous asperities that bristle the surface of most currents. When, however, a deep coating of ashes has subsequently fallen on this surface, its smaller roughnesses are effaced, and the larger protuberances alone shew themselves in the form of small dome-shaped or sub-conical hillocks, which continue, through various crevices, to give a passage to the vapours by which they were at first thrown up. Many such hillocks rising five or six feet above the average level of the surface, existed in the spring of 1823, on the Vesuvian lava currents above-mentioned, and sent forth copious columns of vapour, precisely of the same nature as that of the *Hornitos* of M. de Humboldt; while other fissures intersecting the intervening surface of the small plain formed by the lava on the Pedamentina, gave out similar vapours, presenting on a small scale the identical phenomena for which Malpais has been so long celebrated.†

* See Brieslak. Institutions Geologiques, i. p. 251.

† That the eruptions of Jorullo threw up a prodigious quantity of volcanic ashes, is evident from the fact recorded by M. de Humboldt, that the roofs of the houses at *Queretaro*, 144 miles distant in a straight line, were thickly covered with them!

Where the quantity of ashes covering the lava bed, and mixed up into a paste with rain water, was great, as appears to have been the case on the Malpais, it is probable that numerous hillocks of this kind will have been formed by the intumescence of this semi-liquid substance alone above the fumarole of the lava;‡ and the mobility of parts occasioned by this process, favouring the action of the concretionary forces, probably gave rise to the agglomeration of the clay into the foliated and concentric balls, of which the *hornitos* partly consist. At *Pont du Chateau*, in Auvergne, is an example of even a very coarse calcareo-volcanic conglomerate having assumed this precise variety of concretionary structure; and I suspect, from their being imbedded in the black clay, and consisting of a fine-grained rock very different from the *Doleritic basalt* of Jorullo, that the globular concretions of the Hornitos are not a true basalt, but merely hardened nodular balls of volcanic ashes. They are, in fact, described by M. de Humboldt as *fragile and easily crumbled*, and totally different from the syenitic lava of the current of Jorullo.

It remains only to notice the supplementary facts produced by M. de Humboldt in support of the explanation he adopts of the appearances of Malpais, which I conceive tend much more strongly to confirm the opinion of its being merely the surface of a great bed of lava, which, up to the period of M. de Humboldt's visit, retained much of its internal heat.

These confirmatory facts are,

1. The noise made by the steps of a horse upon the plain.

2. The frequent formation of cracks or fissures across the plain, and the occasional occurrence of partial subsidences.

3. That two rivers, the Cuitimba and San Pedro, lose themselves below the eastern extremity of the plain, and re-appear as hot springs (at 52° cent.) at its western limit.

1. With regard to the first-mentioned circumstance, viz the sound produced when the surface of the plain is struck by the hoofs of a horse, or, I presume, in any other mode of

‡ "The *Hornitos* are *hollow*. When a mule steps upon them they break in."—*Humboldt*.

percussion, it is evidently the same phenomenon of *reverberation*, to which the mimo-phoneutical term *rimbombo* is so well applied in Italy, and which by a vulgar error is often supposed to indicate a great cavity below the spot so resounding when struck. It is perfectly true that the roof of every large cavity does, under certain circumstances, offer the same phenomenon; but the converse of this is by no means true; and to produce this effect it is enough that the soil should be of loose, light, and porous materials, so as to contain numerous small cavities or interstices. Not only the bottom of the crater, but the external slopes of every volcanic cone, and every flat spot however distant from any volcanic orifice, which has a moderate coating of fragmentary scoriæ or volcanic ashes, returns this sound on percussion. Not only the sides of Vesuvius, but the whole surface of the Campagna di Roma, and the Terra di Lavoro, must be suspended over a yawning gulph, if this phenomenon is a sufficient proof of such a position.

But even all *made ground* returns a more or less sonorous reverberation when struck sharply, and the causes which produce this effect are well known to natural philosophers. This sound would therefore be produced in the case of Malpais, as naturally by a superficial coating of volcanic ashes as by any vast cavity, did such indeed exist, underneath.

2. The frequent formation of cracks and fissures across the plain, far from proving the existence of such a subterranean gulph, is a circumstance which accompanies the cooling and consolidation of every bed of lava; and as these crevices are formed only in the lava (contracting as it congeals) it is to be expected that local subsidences must often take place in the coating of volcanic ashes or black clay, immediately above the clefts. The washing of this clay by rains into the fissures of the lava bed beneath, is another probable cause of such subsidences; much more probable I should conceive than the supposition of a natural arched cavity or bubble, four square miles in extent.

3. A further confirmation of the existence of a bed of lava beneath the plain of Malpais is obtained from the disappearance of two rivers beneath its surface; for this accident necessarily results in the instance of all lava currents which have occupied the bed of a river; in consequence of the numerous fissures with which they are penetrated, but particularly of the bed of loose and cellular scoriæ on which they invariably rest. This phenomenon occurs in repeated instances in Auvergne. Wherever a bed of lava fills the bottom of a valley, the river or torrent which drains the valley, dis-

appears below the upper extremity of the lava-bed, and filtering through the interstices of the scoriæ which universally form its substratum, reappears in copious springs at the lower extremity or termination of the lava current. So long as the lava retains a very exalted temperature in its interior, the water percolating beneath it must be proportionately heated; and that this was the case with regard to the lava bed of Malpais at the time of M. de Humboldt's visit was proved by the numerous fumarole on its surface. Hence it was to be expected that the rivers Cuitimba and San Pedro, which find their way beneath it, should have had their temperature raised before they issued again into the air at the opposite extremity of the superinduced lava-bed.

The noise heard on approaching the ear to any of the hillocks (hornitos) resembling that of a cascade, and which is by M. de Humboldt attributed to the rivers flowing through the hollow gulph below, is far more probably owing to the currents of elastic vapour rushing through the fissures by which they find a vent. A similar sound is produced by the rise of carbonic acid through the little cones of the mud volcanos of *Maccaluba* in Sicily; and I have also observed a rushing sound of the same nature to be produced by every powerful *fumarola* of the lava currents of Vesuvius.

M. de Humboldt mentions himself that the heat of the *Hornitos* decreases every year; and I have the authority of Mr. Bullock, junior, who visited the spot a short time back, for the fact that they have now ceased almost entirely to emit vapour, and that the hot springs are reduced to a very low temperature, evidently from the congelation of the subjacent bed of lava. This evidence is absolutely conclusive as to the correctness of the opinions advocated here on the nature of the plain of Malpais.

I have given thus much space to the endeavour to reconcile the phenomena presented by this plain to the ordinary and well known modes of volcanic agency, because the opinion expressed by M. de Humboldt of its surface having been raised by intumescence in the manner of an enormous *bladder or bubble*, (of four square miles in extent!) and covered by an effort of the same extraordinary and incomprehensible nature with thousands of small basaltic cones, each owing to a similar process, has been generally received by Geologists as an ascertained fact, and made the basis of further and still more strange hypotheses for the purpose of explaining the origin of the dome-shaped mountains of frequent occurrence in Trachytic countries, and the still more common conical peaks of basalt.

If, from the reasons adduced above, it appears most probable that the convexity of the plain of Malpais is simply owing to its forming the surface of a massive subjacent bed of lava emitted contemporaneously by the six volcanic cones which rise from its surface, it will be obviously impossible to draw any argument from the formation of the *Hornitos*, none of which exceed nine feet in height, as to that of mountains like the Puy de Dome, Chimboraco or Pichinca, the two latter of which are from 15 to 18,000 feet in height. The theory therefore built on the supposed example of Jorullo must fall to the ground.*

It has been seen in another part of this Essay that the peculiarity of figure assumed occasionally by masses of trachyte and basalt, is easily to be accounted for without having recourse to the agency of unknown and imaginary forces, or indeed any other than those with the operation of which we are thoroughly conversant, and which are fully equal to the purpose.

I trust to be forgiven the apparent presumption of thus calling in question opinions formed by an observer of such acknowledged sagacity and experience as M. de Humboldt, upon facts to which except through his accounts I am necessarily a stranger; no other description of the volcanos of Mexico having, I believe, been made public. I think, however, it must be allowed that the facts of which we have the relation from M. de H. himself, by no means bear out the theory he has proposed to account for them, but tend, on the contrary, one and all, to refer the volcanic eruptions of Jorullo and its vicinity to the same class of phenomena which have been uniformly observed in other localities.

In fact, in the process of argument from effects up to causes, no chain of reasoning can be stronger, no conclusion can be more imperative, than when, as in this instance, we are possessed of a considerable number of facts, all, without one exception, going to support a certain origin, and *that* not an

* The other example adduced by Humboldt for the same purpose, viz. the supposed intumescence of the plains that form themselves on the summit of volcanic cones in place of their craters, is equally inadmissible. It has been already shown, in the body of this work, that the craters are filled by a general law, from the accumulation of fragmentary ejections, and of lava swelling up from below. This will necessarily tend to produce a final convexity of surface; but it would be manifestly absurd to argue from this form the existence of a vaulted cavity below.

imaginary species of phenomenon invented for the occasion, but the same which is observed in its continual operation on other spots to produce the same results, and the only one amongst all known natural processes that is capable of producing them.

I conceive, indeed, that no more effectual service can be rendered to science than the destruction of any one of those glaring theories, which, apparently based upon a few specious facts, and backed by the authority of some great name, are received by the world in general without examination, notwithstanding that they contradict the ordinary march of nature, and consequently throw the extremest perplexity into that of science.

The brilliant theory of the precipitation from one aqueous menstruum of all the crystalline rocks, now beginning to be reduced to its true value, is a striking example of the facility with which the most baseless hypothesis may be imposed on the scientific world as articles of faith, never to be called in question even in thought. Let us trust it will act as a warning for the future.

FINIS.

WILLIAM PHILLIPS, PRINTER.

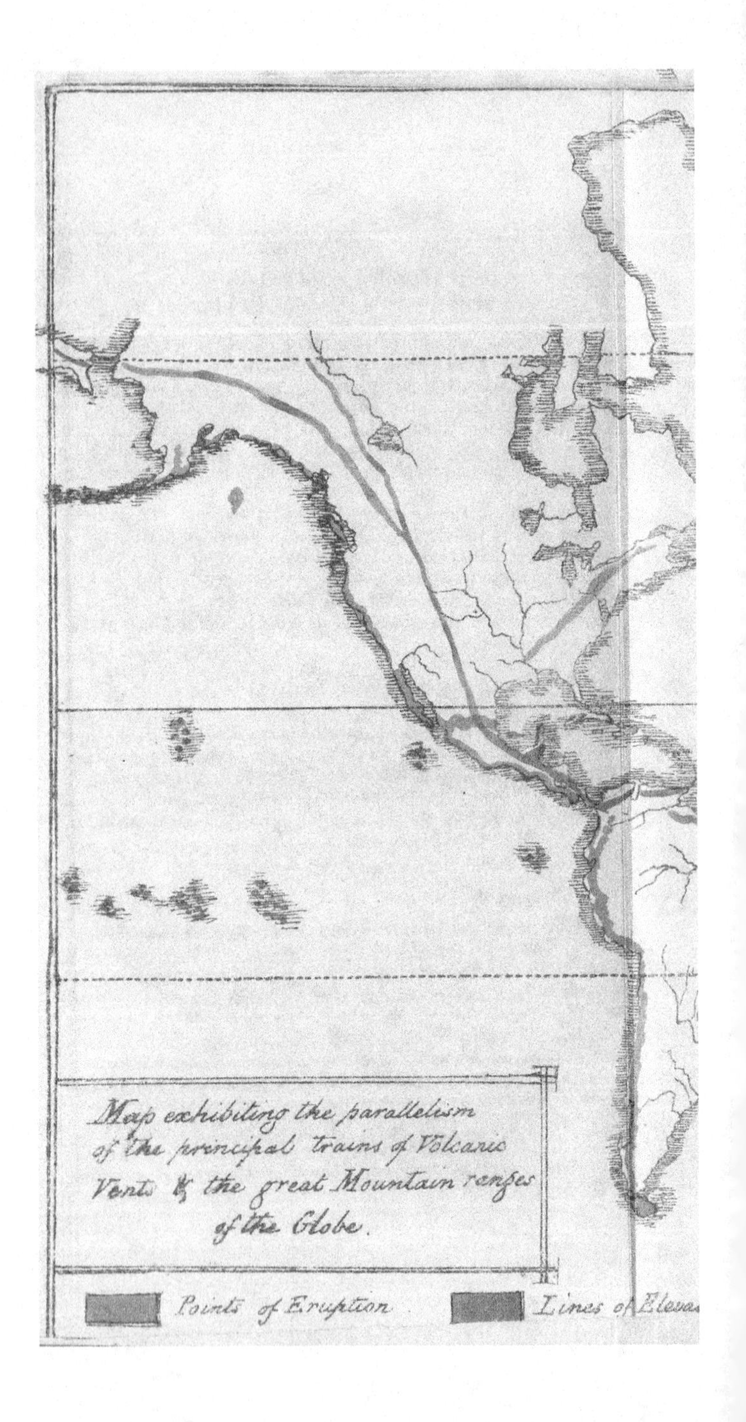
Map exhibiting the parallelism of the principal trains of Volcanic Vents & the great Mountain ranges of the Globe.
Points of Eruption
Lines of Elevat

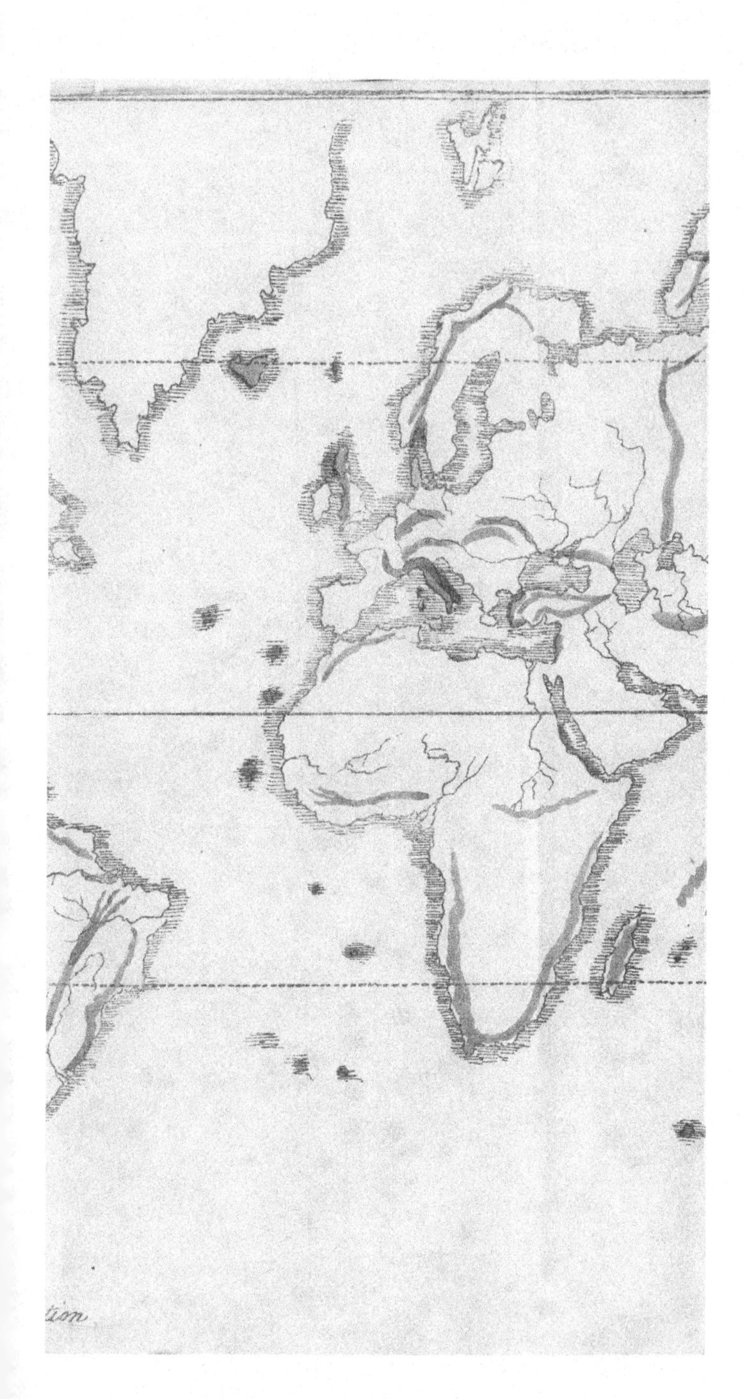
tion

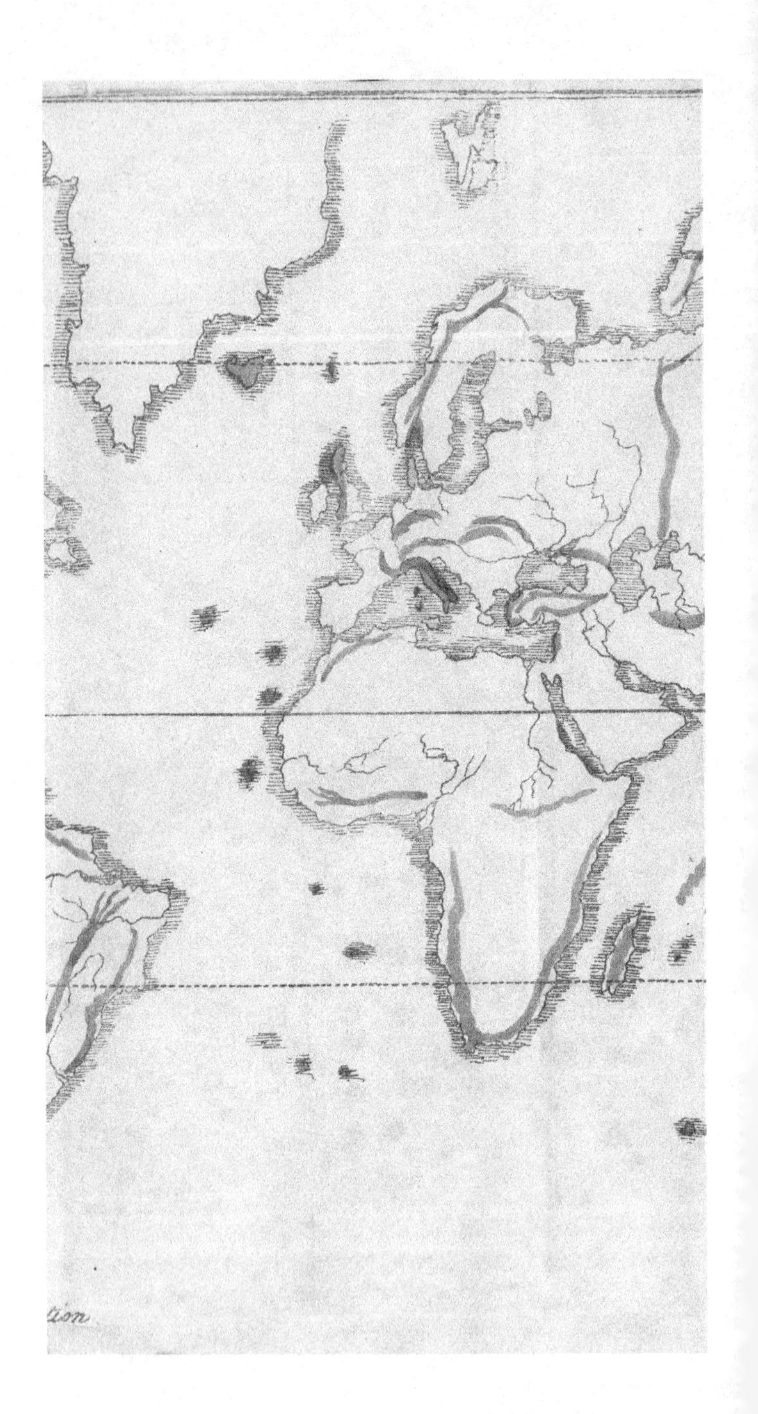

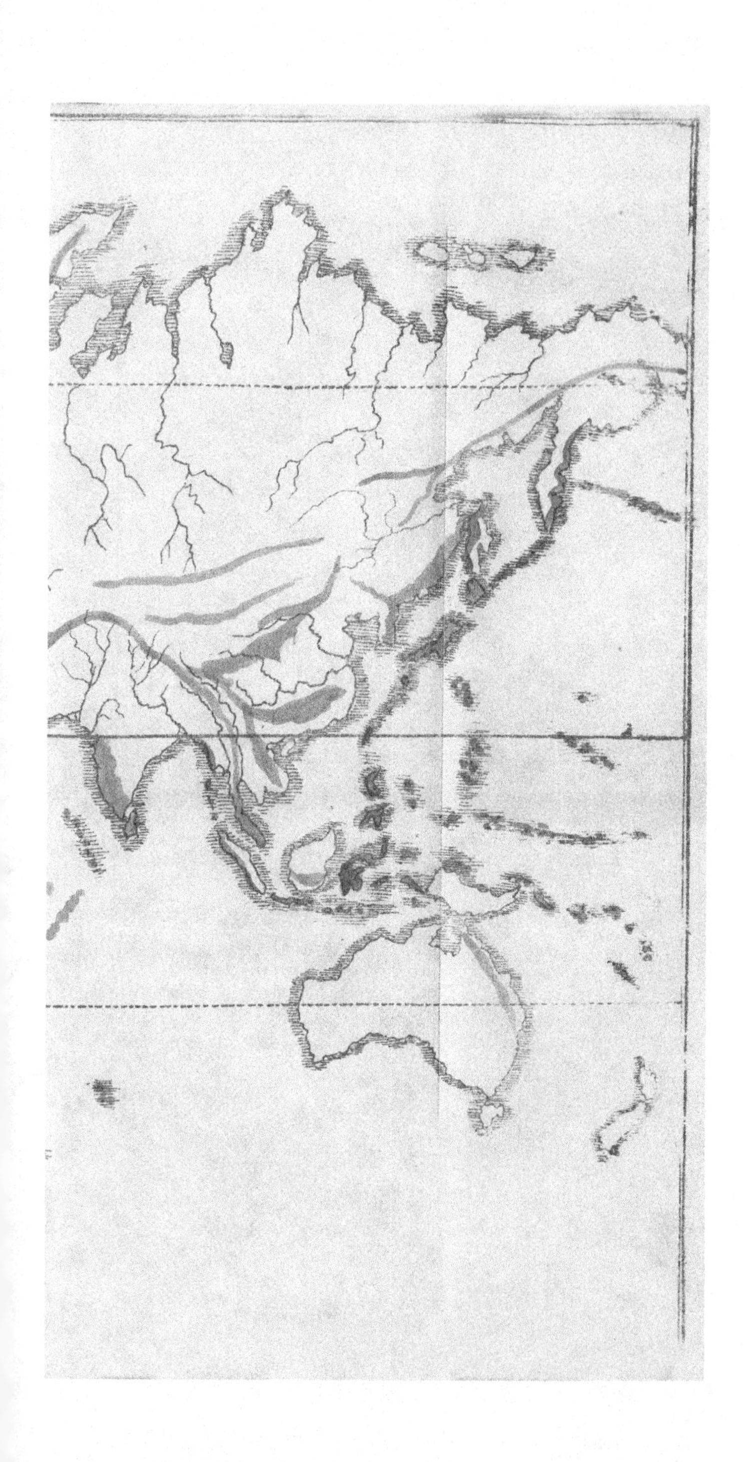

Fig. 2

Granite

Fig. 1

Puys de Montgy
La Taupe

Pourcharet

Montchal

Monjughat

Lamoren

Maison

View of the

Sasso Vernale

Fig. 3

S. Pellegrin

Val di

Sotto i sassi

Bufauro

Vigo

Porphyry

Red Sandstone

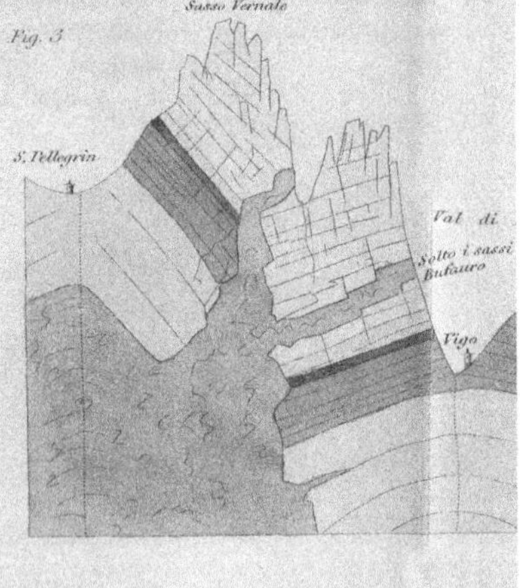

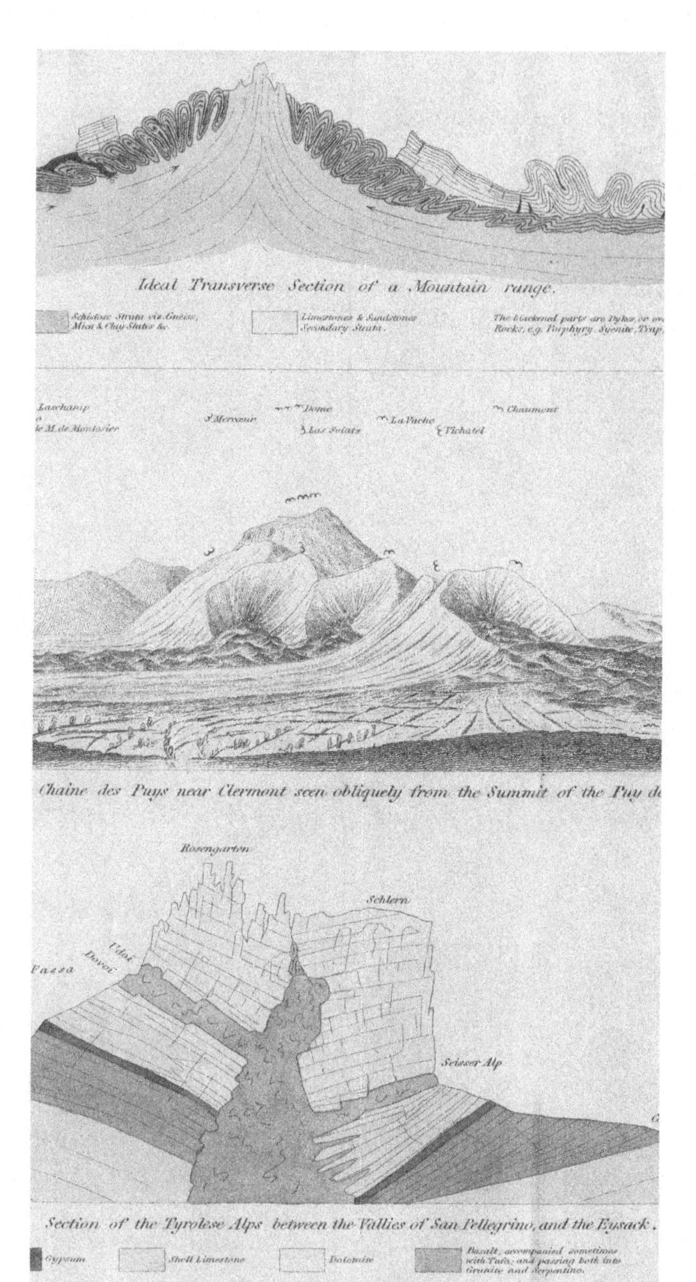

Ideal Transverse Section of a Mountain range.

Chaine des Puys near Clermont seen obliquely from the Summit of the Puy de

Section of the Tyrolese Alps between the Vallies of San Pellegrino, and the Eysack.

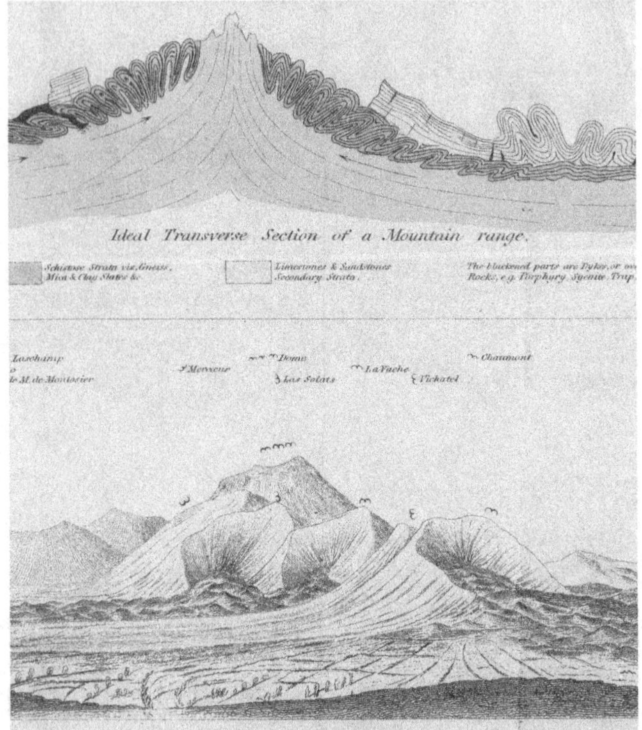

Chaine des Puys near Clermont seen obliquely from the Summit of the Puy de

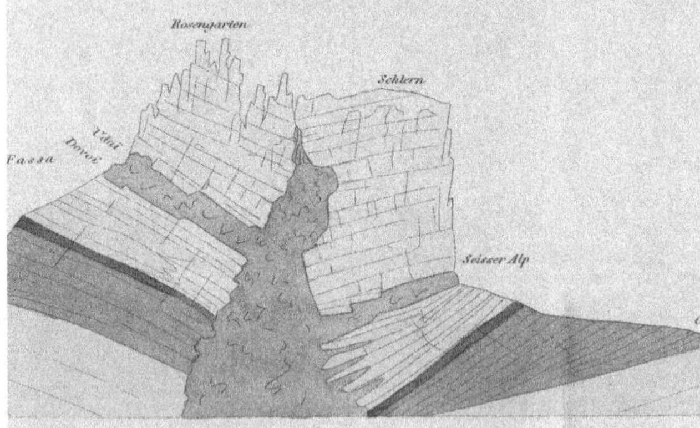

Section of the Tyrolese Alps between the Vallies of San Pellegrino, and the Eysack.

Gypsum | Shell Limestone | Dolomite | Basalt, accompanied sometimes with Tufa; and passing both into Granite and Serpentine.

Published by W. Phillips London 1826.

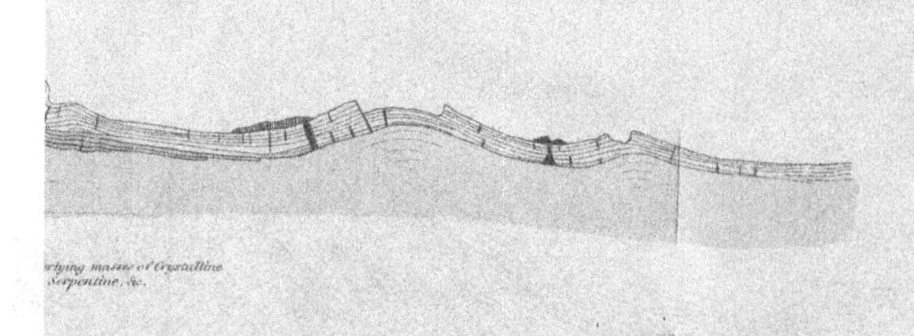

la Rodde.

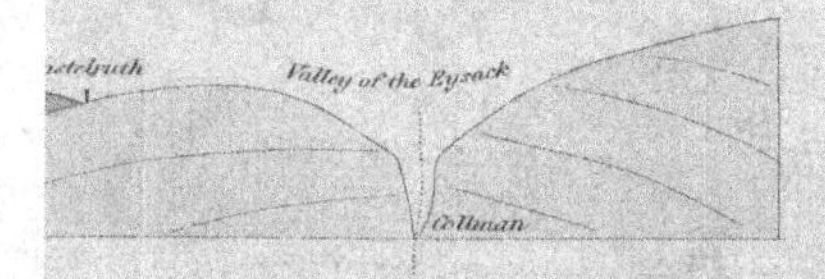

NB. The dotted Lines represent the supposed axes of Elevation.

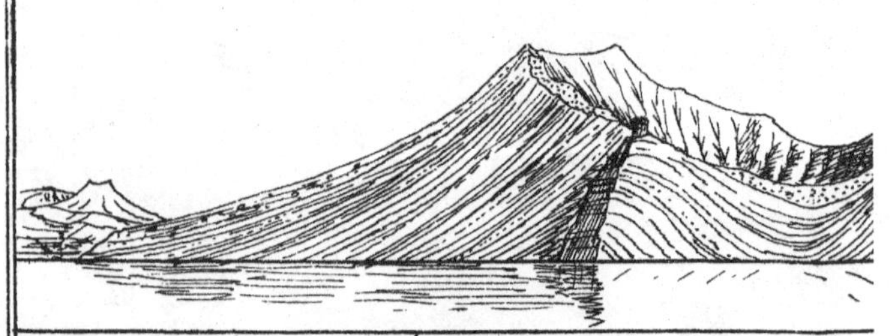

Natural Section of a Volcanic Cone, the Capo di Misen

View of Volcano, one of the Lipari Isles, with Volcanello in

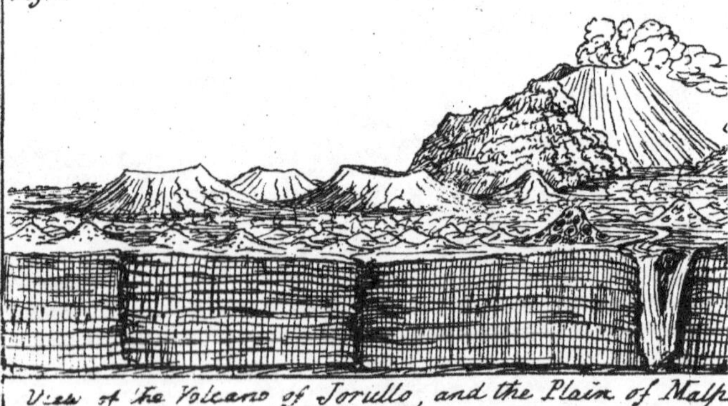

View of the Volcano of Jorullo, and the Plain of Malp

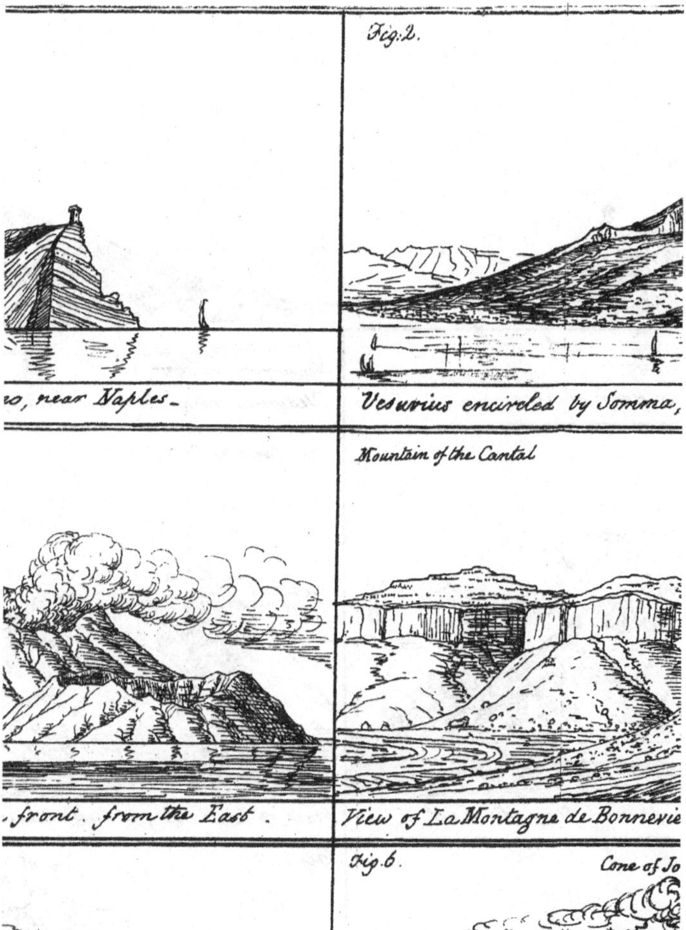

Fig. 2.

...ro, near Naples.

Vesuvius encircled by Somma,

Mountain of the Cantal

...front from the East.

View of La Montagne de Bonnevie...

Fig. 6.

Cone of Jo...

Hornitos

404 toises

Level of the Pl...

Sea Level

...ais covered with Hornitos.

Section of Jorullo and Malp...

Fig: 2.

ro, near Naples.

Vesuvius encircled by Somma,

Mountain of the Cantal

. front. from the East.

View of La Montagne de Bonnevie

Fig. 6.

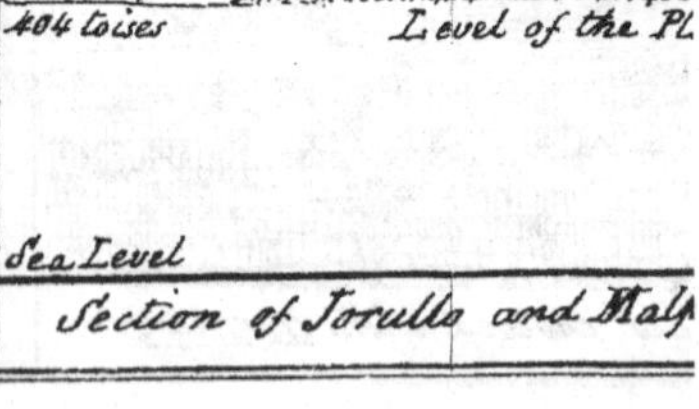

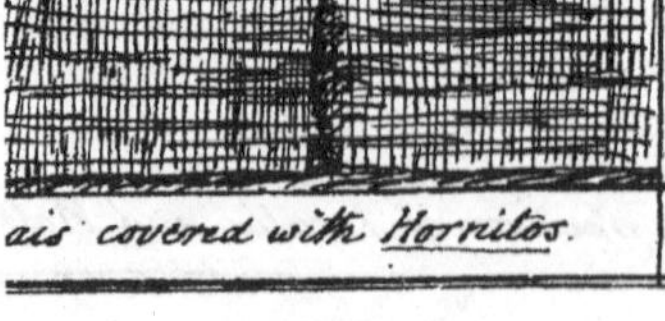

Cone of Jo

Hornitos

404 toises

Level of the Pl

Sea Level

ais covered with Hornitos.

Section of Jorullo and Malp

as seen from the Mountain between Vico and Sorrento.

Fig. 4.

, above the Town of Murat . an immense cluster of Basaltic columns.

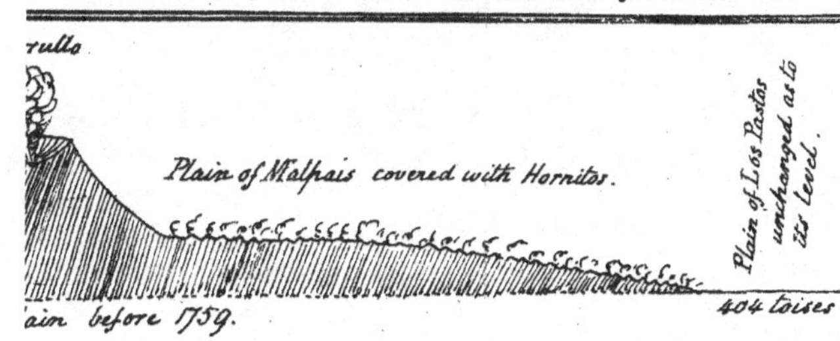

hais, from Humboldts' Relation Historique. Atlas.

Printed in the United States
By Bookmasters